T0324059

Functional
Equations on
Hypergroups

Functional Equations on Hypergroups

László Székelyhidi

University of Debrecen, Hungary

World Scientific

NEW JERSEY • LONDON • SINGAPORE • BEIJING • SHANGHAI • HONG KONG • TAIPEI • CHENNAI

Published by

World Scientific Publishing Co. Pte. Ltd.

5 Toh Tuck Link, Singapore 596224

USA office: 27 Warren Street, Suite 401-402, Hackensack, NJ 07601

UK office: 57 Shelton Street, Covent Garden, London WC2H 9HE

British Library Cataloguing-in-Publication Data
A catalogue record for this book is available from the British Library.

FUNCTIONAL EQUATIONS ON HYPERGROUPS

ISBN 978-981-4407-00-7

Printed in Singapore.

To my wife

Preface

The theory of functional equations is one of the classical fields of mathematics. Functional equation problems arose in different areas from the ancient times both in theory and in applications. In 1966 J. Aczél published his book "Lectures on functional equations and their applications" (see [Acz66]), which is considered the bible of this theory. Although several books, monographs, papers, etc. have since then been published in the field, there is no doubt that this volume is still the most determining reference book. There are other important contributions by J. Aczél and J. Dhombres in "Functional Equations Containing Several Variables" (see [AD89]) and also a basic reference book is due to M. Kuczma ([Kuc09]). The interested reader will find several further references in these books on this wide-ranging field with applications in geometry, geometrical objects, statistics, information theory, utility theory, etc. Here we mention further volumes that have been published more recently, which may convince the reader of the usefulness and effectivity of the diverse methods and application possibilities of the theory of functional equations: [CRC92], [Cor02], [Cze02], [Fel08], [For10], [HIR98], [JS96], [Jár05], [Kan09], [SR98], [SK11], [Szé91].

In the old times functional equations were solved by different ad-hoc – however, ingenious – methods. Anyway, the theory was far from being a compact mathematical discipline in the sense that there were no real general solution methods, no real theories: a good idea would just solved the problem. Later on the situation changed. A pioneer work of A. Járai (see [Jár86], also in [JS96]) – in close connection with Hilbert's Fifth Problem – led to the observation that the strong algebraic character of a functional equation implies important consequences for the analytic behaviour of the solutions: namely, very weak analytic assumptions imply very strong analytic properties. This "regularization theory" was maybe the first important step to build up a coherent theory of functional equations together with its important consequences. The "good old ad-hoc" ideas were replaced by strong theorems and the weight of the theory of functional equations grew similar to that of the theory of differential equations and to other well-respected areas of mathematics. Beside several relevant works of Járai the interested reader will find further references in [JS96]. The comprehensive volume on regularization theory of Járai was published in 2005 [Jár05].

However, another stream started in the 90's with the monograph of the present author (see [Szé91]) emphasizing and introducing the fundamental role of spectral analysis and spectral synthesis in the theory of a special type of functional equations: the so-called convolution type functional equations. Convolution type functional equations are actually integral equations and it turns out that a major part of the so-called "classical" equations belongs to, or can be reduced to this type. In the monograph [Szé91] the author offers a general method for the solution of convolution type systems of functional equations. The essence of the method is that first the "basic building blocks" of the solution space of the functional equation should be found – these are the so-called "exponential monomials" – and then – in case of spectral synthesis – the linear combinations of these basic solutions will form a dense set in the solution space, that is, they characterize the solution space. It happens, or not, the exponential monomial solutions play a very special and important role in the solution process.

It turns out that several ideas of this type can be adopted to a more delicate situation: to the situation of hypergroups. The concept of DJS–hypergroup, which we shall use here (according to the initials of C. F. Dunkl, R. I. Jewett and R. Spector) is due to R. Lasser (see e.g. [Ros98], [BH95]), [Las83]. One can realize a hypergroup like the convolution structure of some measure algebra over a group, but the group structure has been neglected. If x, y are elements of a hypergroup, then the notation $x * y$ has a symbolic meaning only: it does not represent an element of the hypergroup, just a kind of "blurred product". In the group-case $x \cdot y$ is a well-defined element of the underlying structure, which also can be considered as a measure $\mu_{x,y}$ with the property that for any set B the value $\mu_{x,y}(B)$ is equal to 1 if $x \cdot y$ belongs to B and it is equal to 0, if $x \cdot y$ does not belong to B. Hence this is exactly the point mass concentrated at $x \cdot y$. However, in the case of a hypergroup, $x * y$ denotes a measure, actually $\delta_x * \delta_y$, which is not necessarily a point mass and $x * y(B)$ represents the "probability" of the event that the "product" $x * y$ belongs to B.

Anyway, using $x * y$ one can introduce translation operators on hypergroups, which makes it possible to set up a theory of harmonic analysis. The interested reader will find further details and references in the fundamental work of W. R. Bloom and H. Heyer ([BH95]). The theory of hypergroups has been a developing field, where ideas from different areas of mathematics can be utilized to obtain general results, which may help better understanding also in the classical situation. The interested reader will find further references and better insight in the papers [Las83], [Ros77], [Ros98], [AC11], [KPC10], [OEBG10], [Sam10], [Hey09], [LBPS09], [LBS09], [Mur08], [BH08], [DK07], [EL07], [Pav07c], [LOR07], [Mur07], [Młlo06], [HK06], [BR05], [Las05], [FLS05], [SW03], [BBM02], [Gha02], [Las02], [BR02], [Kum01], [Hin00], [NI00], [GS00], [Ros99], [Pav99a], [Pav99b], [RV99], [CV99], [CS98], [Tri98], [Sch98], [RAL⁺98], [Par97], [Tri97], [OW96], [Wil95], [Rös95b], [Ros95c], [Hey95], [Ché95], [BK95], [CGS95].

As soon as translation operators appear on hypergroups a wide range of machinery can be adopted from the group-case. Nevertheless, the classical group-methods can be applied only restrictively: the special situation does not make it possible to "copy" the well-known classical methods. However, there are some distinguished function classes, like additive functions, exponential functions, or more generally, exponential polynomials, which play a vital role on both groups and hypergroups. Another class is represented by the so-called moment functions, which are extremely important in the different applications of hypergroups in probability theory and statistics. For more about these function classes the interested reader will find detailed information in [BH95], [Szé91], [Gal98], [Ros98], [Zeu92], [OS05], [OS04], [OS08], [Szé06c], [Gal97].

The appearance of translation operators enables us to utilize a very effective method of studying functional equations and systems of functional equations on hypergroups. Namely, it turns out that some of the methods of spectral analysis and spectral synthesis can be adopted and used in the hypergroup-situation. The present author has recently published a volume about the applications of spectral analysis and spectral synthesis on different structures ([Szé06a]). In that monograph the interested reader will find detailed information about spectral analysis, spectral synthesis and their use in the theory of functional equations. However, it turns out that several new ideas and methods can be transformed into the hypergroup-situation, which may enrich both fields: the theory of functional equations and the theory of hypergroups. To present the fruitful consequences of this delicate "marriage" was one of the main purpose to write this volume. The interested reader will find further results and references on these connections in [Ros98], [Gal98], [Las83], [Zeu92], [AC11], [RZ11], [Vaj10b], [KPC10], [OEBG10], [AK10], [HP10], [Hey09], [Las09a], [LBPS09], [NS09], [FK08], [Azi08], [Mur08], [BH08], [Ami07], [Pav07b], [HL07], [Pav07c], [Mur07], [Mło06], [HK06], [Men05], [BR05], [Las05], [FLS05], [Pav04], [SW03], [NI03], [HL03], [Wil02], [Gha02], [GT02b], [Gal02], [Las02], [GT02a], [Kum01], [NI01], [Hin00], [NI00], [GS00], [FL00], [Ros99], [Pav99a], [Pav99b], [NI99], [RV99], [CV99], [GS99], [Gal99], [Ren98], [NS98], [CS98], [Geb98], [Zeu98], [Tri98], [SW98], [Sch98], [Par97], [Pav97], [Wil97], [Zeu97], [KS97], [Flo96], [BH96], [Ren96], [BR96], [OW96], [Ehr96], [Sin96], [Zeu95b], [Wil95], [Rös95a], [CS95b], [Rös95b], [Her95], [HV95], [OEBB95], [Zeu95a], [Zeu95a], [Voi95], [Voi95], [CS95a], [BK95], [CGS95], [Szw95], [Han94], [RX94], [Zeu94], [Las94], [Voi93], [Hey93], [RX93], [LOR07], [BR02], [RAL$^+$98].

In what follows we try to give a brief overlook about the structure of this booklet, the fundamental methods and the main results.

This Preface is followed by an Introduction in which we summarize the most important concepts concerning hypergroups. Some of these concepts are analogous to those of the ones in the group-case, but sometimes we meet basic differences. However, the concepts of additive functions, exponential functions, exponential monomials and exponential polynomials are introduced here and the relation to the corresponding group-case concepts is presented. Another important function class is the class of moment functions, mentioned above, which plays a very important role in the applications of hypergroups in probability theory and statistics. The interested reader should refer to [BH95], [Gal98], [Ros98], [Zeu92] and the references included in these works.

The Introduction is also devoted to present those analytic methods, which are very effective in the group case to prove strong regularity of solutions of functional equations assuming their weak regularity, only. The basic tools are Haar measure and invariant means. Here we tried to present a unified, nonstandard treatment of these basic analytic tools.

The next two chapters are devoted to the study of functional equations on a very important type of hypergroups: polynomial hypergroups in one variable and polynomial hypergroups in several variables. Here we give the complete description of those basic functional classes mentioned above, which play an important role in the applications: additive functions, exponential functions and exponential polynomials. Although the two chapters are very closely related to each other, the author's idea was to separate the consideration of polynomial hypergroups in one variable and in several variables. The reason is that sometimes the methods are basically different and this kind of "separation" may help better understanding. For detailed information about polynomial hypergroups the reader should refer to [Las83], [Szé04], [Vog87], [HHL10], [Las09b], [Las09a], [Szé08], [BH08], [Las07], [LOR07], [Młoo6], [HL03], [BR02], [Tri00], [GS99], [Ehr96], [Zeu95b], [CS95a], [Szw95].

In Chapter 5 a new type of hypergroups appears: the so-called Sturm–Liouville hypergroups, which play a fundamental role in the theory of hypergroups, differential equations and initial value problems. It turns out that some of the above mentioned important function classes can be introduced, studied and characterized on these types of hypergroups. For more information about general and special Sturm–Liouville-hypergroups see e.g. [Ché95], [Szé06b], [OS08], [Vaj10b], [Vaj10a], [DK07], [Ole01], [Ché95], [Szé06b], [OS08], [Zeu89], [Ma08], [Tri05a], [Tri05b], [BBM02], [BX00b], [BX00a], [NRT98], [JT98], [BX98a], [BX98b], [BX97], [LT95], [BX95].

Chapter 6 contains three sections on the so-called two-point support hypergroups. Here we illustrate how the advanced methods of the theory of functional equations can be utilized to characterize some basic function classes on different types of hypergroups. We exhibit an example for two-point hypergroups of compact and of noncompact type, moreover another one, the so-called cosh *hypergroup*, which has been studied by H. Zeuner in [Zeu89].

Chapters 7 and 8 are – in some sense – the heart of this book: spectral analysis and spectral synthesis on different special types of hypergroups. Spectral analysis and spectral synthesis have become effective tools in functional equations recently. The classical roots go back to harmonic analysis and Fourier series. The abstract background can be found in [Loo53]. Basic knowledge and results on classical spectral theory of linear operators and spectral synthesis can be found in [Ben75], [Beu48], [Hel52], [Ris49], [Szé02], [Szé06a], [MA50], [Mal54], [Vog87], [DS88a], [DS88b], [DS88c], [Lef58], [Mal59], [Sch48], [Hel83], [HR63]. Studying harmonic analysis on hypergroups is possible because of the presence of translation operators. The fundamentals of this theory are presented in [BH95]. The use of harmonic analysis and synthesis in the theory of functional equations was invented in [Szé91]. However, the group-methods are not always easy to adopt in the hypergroup situation: sometimes new ideas are needed. Nevertheless, here the field is open as we were able to prove spectral analysis and spectral synthesis theorems for a restricted class of hypergroups, only. However, we hope that the applications we present here will convince the reader that further investigations in this area may lead to interesting and useful results – both for functional equationists and for hypergroup experts. We just mention that – as it is clear from the results of Sections 2.2, 3.2 and 4.2 – it is a nontrivial problem on how to define exponential monomials on arbitrary (commutative) hypergroups.

Chapter 9 is devoted to a classical problem of probability theory: the moment problem (see [Akh65], [Sti94]). We formulate the problem on commutative hypergroups and solve the uniqueness in the case of polynomial hypergroups in a single variable and of Sturm–Liouville-hypergroups.

In Chapters 10 and 11 we collected diverse applications of spectral analysis, spectral synthesis and other methods. These applications are illustrated on different classical and non-classical functional equations. For instance, in Chapter 11 the reader meets a new theory of difference equations on hypergroups – at least the basic and far-leading ideas.

The closing Chapter 12 is devoted to a special field of functional equations: stability theory. Since the pioneer talk of Stanislaw Ulam in 1940 presented to the audience of the Mathematics Club of the University of Wisconsin the door has been opened to a completely new world of investigations: stability became a central problem in the theory of functional equations (see [Cor02], [HIR98], [Hey93], [Cze02]). Here we make an attempt to outline some possible ways climbing these mountains on hypergroups.

This volume is completed with a list of references and a subject index.

We hope that the present work is able to represent faithfully the possibilities of connecting functional equation problems with those coming from the theory of hypergroups. We are convinced that both areas will profit from a "come together" of this type. This volume is written for those who have open eyes for both meadows, who have open ears for both concerts and who dare to enter a new world of ideas, a new world of methods – and, sometimes, a new world of unexpected difficulties.

The author is indebted to all those who helped in this work to become complete. Finally, I would like to express my special thanks to my students, Ágota Orosz and László Vajday, who did their best, who provided the newest results and without whose contribution this work could not have been accomplished.

László Székelyhidi
2012

Contents

Chapter 1

Introduction

1.1 Basic concepts and facts

The major part of this section is taken from [BH95]. The concept of DJS–hypergroup (according to the initials of C. F. Dunkl, R. I. Jewett and R. Spector) depends on a set of axioms which can be formulated in several different ways. The way of formulating these axioms we follow here is due to R. Lasser (see e.g. [BH95], [Ros98]). One begins with a locally compact Hausdorff space K and with the space $\mathcal{C}_c(K)$ of all compactly supported complex valued functions on the space K. The space $\mathcal{C}_c(K)$ will be topologized as the *inductive limit* of the spaces

$$\mathcal{C}_E(K) = \{f \in \mathcal{C}_c(K) \ : \ supp\,(F) \subseteq E\}\,,$$

where E is a compact subset of K carrying the uniform topology. A (complex) *Radon measure* μ is a *continuous linear functional* on $\mathcal{C}_c(K)$. Thus, for every compact subset E in K there exists a constant α_E such that $|\mu(f)| \leq \alpha_E ||f||_\infty$ for all f in $\mathcal{C}_E(K)$. The set of Radon measures on K will be denoted by $\mathcal{M}(K)$. For every μ in $\mathcal{M}(K)$ we write

$$||\mu|| = \sup\{|\mu(f)| \ : \ f \in \mathcal{C}_c(K), ||f||_\infty \leq 1\}.$$

A measure μ is said to be *bounded*, if $||\mu|| < +\infty$. In addition, μ is called a *probability measure*, if μ is nonnegative and $||\mu|| = 1$. The set of all bounded measures, the set of all compactly supported measures, the set of all probability measures and the set of all probability measures with compact support in $\mathcal{M}(K)$ will be denoted by $\mathcal{M}_b(K)$, $\mathcal{M}_c(K)$, $\mathcal{M}_1(K)$

and $\mathcal{M}_{1,c}(K)$, respectively. The point mass concentrated at x is denoted by δ_x. Via integration theory we are able to consider measures as functions on the σ-algebra $\mathcal{B}(K)$ of *Borel subsets* of K and we use the notation $\int_K f \, d\mu$ rather than $\mu(f)$ even when either is possible. We use the notation $\mathcal{M}_+(K)$ for the set of positive measures on the σ-algebra $\mathcal{B}(K)$ that means, for measures which take values in $[0, +\infty]$.

Now we formulate the first part of the axioms. Suppose that we have the following:

(H^*) There is a continuous mapping $(x, y) \mapsto \delta_x * \delta_y$ from $K \times K$ into $\mathcal{M}_{1,c}(K)$. This mapping is called *convolution*.

$(H^\vee)$ There is an involutive homeomorphism $x \mapsto x^\vee$ from K to K. This mapping is called *involution*.

(He) There is a fixed element e in K. This element is called *identity*.

Identifying x by δ_x the mapping in (H^*) has a unique extension to a continuous bilinear mapping from $\mathcal{M}_b(K) \times \mathcal{M}_b(K)$ to $\mathcal{M}_b(K)$. The involution on K extends to a continuous involution on $\mathcal{M}_b(K)$. Convolution maps $\mathcal{M}_1(K) \times \mathcal{M}_1(K)$ into $\mathcal{M}_1(K)$ and involution maps $\mathcal{M}_1(K)$ onto $\mathcal{M}_1(K)$. Then a *DJS–hypergroup*, or simply a *hypergroup* is a quadruple $(K, *, \vee, e)$ satisfying the following axioms: for each x, y, z in K we have

(H1) $\delta_x * (\delta_y * \delta_z) = (\delta_x * \delta_y) * \delta_z$,

(H2) $(\delta_x * \delta_y)^\vee = \delta_{y^\vee} * \delta_{x^\vee}$,

(H3) $\delta_x * \delta_e = \delta_e * \delta_x = \delta_x$,

(H4) e is in the support of $\delta_x * \delta_{y^\vee}$ if and only if $x = y$,

(H5) the mapping $(x, y) \mapsto supp\,(\delta_x * \delta_y)$ from $K \times K$ into the space of nonvoid compact subsets of K is continuous, the latter being endowed with the Michael topology (see [BH95]).

For any measures μ, ν in $\mathcal{M}_b(K)$ obviously $\mu * \nu$ denotes their convolution and $\mu^\vee$ denotes the involution of μ. With these operations $\mathcal{M}_b(K)$ is an *algebra with involution*. If the topology of K is discrete, then we call the hypergroup *discrete*. In case of discrete hypergroups the above axioms have a simpler form. As in this book we frequently will focus on discrete hypergroups, here we present a set of axioms for these types of hypergroups. Clearly, in the discrete case we can simply forget about the topological requirements in the previous axioms to get a purely algebraic system.

Let K be a set and suppose that the following properties are satisfied:

(D^*) There is a mapping $(x, y) \mapsto \delta_x * \delta_y$ from $K \times K$ into $\mathcal{M}_{1,c}(K)$, the space of all finitely supported probability measures on K. This mapping is called *convolution*.

$(D^\vee)$ There is an involutive bijection $x \mapsto x^\vee$ from K to K. This mapping is called *involution*.

(De) There is a fixed element e in K. This element is called *identity*.

Identifying x by δ_x as above and extending convolution and involution, a *discrete DJS–hypergroup* is a quadruple $(K, *, \vee, e)$ satisfying the following axioms: for each x, y, z in K we have

(D1) $\delta_x * (\delta_y * \delta_z) = (\delta_x * \delta_y) * \delta_z$,

(D2) $(\delta_x * \delta_y)^\vee = \delta_{y^\vee} * \delta_{x^\vee}$,

(D3) $\delta_x * \delta_e = \delta_e * \delta_x = \delta_x$,

(D4) e is in the support of $\delta_x * \delta_{y^\vee}$ if and only if $x = y$.

If $\delta_x * \delta_y = \delta_y * \delta_x$ holds for all x, y in K, then we call the hypergroup *commutative*. If $x^\vee = x$ holds for all x in K, then we call the hypergroup *Hermitian*. By (H2), any Hermitian hypergroup is commutative. In any case we have $e^\vee = e$. For instance, if $K = G$ is a locally compact Hausdorff–group, $\delta_x * \delta_y = \delta_{xy}$ for all x, y in K, $x^\vee$ is the inverse of x and e is the identity of G, then we obviously have a hypergroup $(K, *, \vee, e)$, which is commutative if and only if the group G is commutative. However, not every hypergroup originates in this way.

The simplest hypergroup is obviously the trivial one, consisting of a singleton. The next simplest hypergroup structure can be introduced on a set consisting of two elements. Now we describe all hypergroups of this type. Let $K = \{0, 1\}$. Clearly, the only Hausdorff topology on K is the discrete one. We specify $e = 0$ as the identity element. In this case the only involution satisfying the above axioms is the identity, that is, $0^\vee = 0$ and $1^\vee = 1$. Consequently, we have a Hermitian hypergroup, which is necessarily commutative. Now we have to define the four possible products $\delta_0 * \delta_0$, $\delta_0 * \delta_1$, $\delta_1 * \delta_0$ and $\delta_1 * \delta_1$. As δ_0 is the identity, the first three products are uniquely determined and the fourth one must have the form

$$\delta_1 * \delta_1 = \theta \cdot \delta_0 + (1 - \theta) \cdot \delta_1$$

with some number θ satisfying $0 \le \theta \le 1$. It turns out that $\theta \ne 0$, as a consequence of (D4). We shall denote this hypergroup by $D(\theta)$. It is clear that in this way we have a complete description of all possible hypergroup structures on a set consisting of two elements. Observe that in the case $\theta = 1$ we have a group isomorphic to $\mathbb{Z}_2$, the integers modulo 2, in any other case the resulting structure is not a group.

If K is any hypergroup and H is an arbitrary set, then for the function $f : K \to H$ we define $f^\vee$ by the formula

$$f^\vee(x) = f(x^\vee)$$

for each x in K. Obviously $\left(f^\vee\right)^\vee = f$. Any measure μ in $\mathcal{M}_b(K)$ satisfies

$$\mu^\vee(f) = \mu(f^\vee)$$

for any bounded Borel function $f : K \to \mathbb{C}$.

Let K be any hypergroup. Then, for each x, y in K the measure $\delta_x * \delta_y$ is a compactly supported probability measure on K, which makes the measurable space $(K, \mathcal{B}(K), \delta_x * \delta_y)$ a *probability space*. Any function $f : K \mapsto \mathbb{C}$, which is $\delta_x * \delta_y$-measurable, can be considered as a *random variable* on this probability space. In particular, any continuous complex valued function on K is a random variable with respect to any measure of the form $\delta_x * \delta_y$. Clearly, any f is integrable with respect to each δ_x and its *expectation* is

$$E_x(f) = \int f \, d\delta_x = f(x) \, ,$$

hence it seems to be reasonable to define the "value" of f at δ_x as $f(x)$. This can be extended to any probability measure μ on K by defining

$$f(\mu) = E_\mu(f) = \int f\,d\mu\,,$$

whenever f is integrable with respect to μ. In particular,

$$f(\delta_x * \delta_y) = \int f\,d(\delta_x * \delta_y)\,,$$

whenever f is integrable with respect to $\delta_x * \delta_y$. In this case we shall use the suggestive notation $f(x * y)$ for $f(\delta_x * \delta_y)$. Actually, in any hypergroup K we identify x by δ_x.

Here we call the attention to the fact that $f(x*y)$ has no meaning on its own, because $x * y$ is in general not an element of K, hence f is not defined at $x*y$. The expression $x*y$ denotes a kind of "indistinct" product. If B is a Borel subset of K, then $\delta_x * \delta_y(B)$ expresses the probability of the event that this "blurred product" of x and y belongs to the set B. In the special case of groups this probability is 1 if B contains xy and is 0 otherwise, that is, exactly $\delta_{xy}(B)$.

We define the *right translation operator* τ_y by the element y in K according to the formula

$$\tau_y f(x) = \int_K f\,d(\delta_x * \delta_y)$$

for any f integrable with respect to $\delta_x * \delta_y$. In particular, τ_y is defined for any continuous complex valued function on K. Similarly, we can define *left translation operators*, denoted by $_y\tau$. In general, one uses the above notation

$$f(x * y) = \int_K f\,d(\delta_x * \delta_y)\,,$$

for each x, y in K. Obviously, in case of commutative hypergroups the simple term *translation operator* is used. The function $\tau_y f$ is *the translate of f by y.*

Convolution of functions and measures is defined in the following way: for any measure μ in $\mathcal{M}_b(K)$ and for any continuous bounded function $f : K \to \mathbb{C}$ we let

$$f * \mu(x) = \int_K f(x * y^{\vee}) \, d\mu(y)$$

for each x in K. Then $f * \mu$ is a continuous bounded function on K. For more details see [BH95].

1.2   Convolution of subsets

The convolution of measures makes it possible to introduce the *convolution of subsets* as follows. Let K be a hypergroup and let A, B be arbitrary subsets in K. The *convolution of A and B* is the set

$$A * B = \bigcup \{ supp \, (\delta_x * \delta_y) : x \in A, y \in B \}. \tag{1.1}$$

It is easy to see that $A * (B * C) = (A * B) * C$ holds for any subsets A, B, C of K. If x is in K and B is a subset of K, then we write $x * B$ for the set $\{x\} * B$.

For any subset A of K we denote by $A^{\vee}$ the set $\{a^{\vee} : a \in A\}$. A subset A is called *symmetric*, if $A = A^{\vee}$. For any subsets A, B of K we have that e is in $A^{\vee} \cap B$ if and only if $A \cap B \neq \emptyset$. Also $(A * B)^{\vee} = B^{\vee} * A^{\vee}$. The following statement will be very useful in the sequel (see [BH95]).

Theorem 1.1. *Let K be a hypergroup and let A, B, C be arbitrary subsets of K. Then we have*

$$(A * B) \cap C \neq \emptyset \qquad \textit{if and only if} \qquad B \cap (A^{\vee} * C) \neq \emptyset. \tag{1.2}$$

Proof. $(A * B) \cap C \neq \emptyset$ if and only if

$$e \in (A * B)^{\vee} * C = B^{\vee} * (A^{\vee} * C),$$

which holds if and only if $B \cap (A^{\vee} * C) \neq \emptyset$. $\qquad\qquad \square$

Another simple technical statement is included in the following theorem.

Theorem 1.2. *Let K be a commutative hypergroup, let B be a symmetric subset of K and let x, y be in K. Then $x \in y * B$ if and only if $y \in x * B$.*

Proof. Suppose that $x \in y * B$ then $\{x\} \cap (y * B) \neq \emptyset$. We apply Theorem 1.1 with the choice $A = B^\vee = B$, $C = \{y\}$ and $B = \{x\}$. Then we have $(x * B) \cap \{y\} \neq \emptyset$, hence $y \in x * B$. $\qquad\qquad\square$

One can prove easily that if A is an arbitrary and B is an open subset in K then $A * B$ and $B * A$ are open, further if A is compact and B is closed, then $A * B$ and $B * A$ are closed. In addition, we have

$$supp\,(\delta_x * \delta_y) = \{x\} * \{y\} \tag{1.3}$$

holds for each x, y in K. For more about the elementary properties of convolution of sets see [BH95] and [Jew75].

1.3 Invariant means on hypergroups

Invariant linear functionals and operators are very useful tools on topological groups. In particular, invariant means can be used to prove stability theorems for functional equations. For this technique see e.g. [Szé00]. Due to the presence of translation operators one can easily formulate the concept of invariant mean on hypergroups, as well.

Let K be a locally compact Hausdorff space and let $\mathcal{CB}(K)$ denote the Banach space of all continuous bounded complex valued functions on K equipped with the *supremum norm*. A function $M : \mathcal{CB}(K) \to \mathbb{C}$ is called a *mean* on K if it satisfies the following properties: for each f, g in $\mathcal{B}(K)$ and complex number λ we have

(1) $M(f + g) = M(f) + M(g)$,

(2) $M(\lambda \cdot f) = \lambda \cdot M(f)$,

(3) $M(\overline{f}) = \overline{M(f)}$,

(4) $\inf f \leq M(f) \leq \sup f$, if f is real valued.

It follows that M is a linear functional on $\mathcal{CB}(K)$ having the properties that M is real valued for real valued functions, $M(1) = 1$ and $M(f) \geq 0$

if f is real valued and $f \geq 0$, hence $M(f) \leq M(g)$ holds for any f, g real valued functions with $f \leq g$. On the other hand, these properties actually characterize the means, as it is shown in the following theorem.

Theorem 1.3. *Let M be a linear functional on $CB(K)$, which takes non-negative real values on nonnegative real valued functions and $M(1) = 1$. Then M is a mean on K.*

Proof.   First of all any real valued function f in $CB(K)$ can be written in the form $f = f^+ - f^-$, where f^+ and f^- denote the positive and negative part of f, respectively, that is,

$$f^+ = \frac{1}{2}(|f| + f), \qquad f^- = \frac{1}{2}(|f| - f).$$

Clearly f^+ and f^- are real valued nonnegative functions. It follows that

$$M(f) = M(f^+) - M(f^-)$$

is a real number, hence M takes real values on real valued functions. Moreover, if f, g are real valued in $CB(K)$ and $f \leq g$, then $g - f$ is nonnegative, hence

$$M(g) - M(f) = M(g - f) \geq 0$$

and we have $M(f) \leq M(g)$.

Now let f be given in $CB(K)$ and we write f in the form $f = a + ib$ with some real valued functions a, b in $CB(K)$. Then $\overline{f} = a - ib$ and

$$M(f) = M(a + ib) = M(a) + i M(b),$$

further

$$M(\overline{f}) = M(a - ib) = M(a) - i M(b).$$

As $M(a)$ and $M(b)$ are real numbers, hence $M(\overline{f}) = \overline{M(f)}$.

By the linearity and $M(1) = 1$ it follows $M(c) = c$ for any constant function c. Hence we have for any real valued f in $\mathcal{CB}(K)$

$$\inf f = M(\inf f) \leq M(f) \leq M(\sup f) = \sup f$$

and the theorem is proved. □

Sometimes we write $M_x\big(f(x)\big)$ for $M(f)$ if we wish to emphasize that M is applied for a function of x.

Suppose now that K is a hypergroup. The mean M on the hypergroup K is called a *right invariant mean*, if

$$M(f) = M_x\big(f(x * y)\big)$$

holds for each f in $\mathcal{CB}(K)$. Clearly $f(x * y)$ is defined for each x, y in K, as f is integrable with respect to $\delta_x * \delta_y$. Right invariance of a mean expresses the fact that this mean has the same value on any right translate of a given function. Similarly, we define *left invariant means* by the analogous property. If a mean is simultaneously right invariant and left invariant, then we call it an *invariant mean*.

A hypergroup is called *right amenable*, *left amenable* or *amenable* if there exists a right invariant mean, a left invariant mean or an invariant mean on K, respectively. The problem of characterization of amenable hypergroups is open, which is nontrivial even in the group case. For more about invariant means, amenable semigroups and groups see [Gre69]. However, an invariant mean always exists on commutative hypergroups, as it is shown in Theorem 1.5 (see also [Lau83]). The proof is based on the Markov–Kakutani fixed-point theorem (see e.g. [DS88a], [Con90]).

Theorem 1.4. *(Markov–Kakutani) Let C be a compact convex set in a locally convex Hausdorff topological vector space. Then every commuting family of continuous endomorphisms on C has a common fixed point.*

Using this theorem we can prove the following result.

Theorem 1.5. *Every commutative hypergroup is amenable.*

Proof. Let K be the underlying locally compact topological space of a commutative hypergroup and we consider the dual space E of the Banach space $\mathcal{CB}(K)$ equipped with the weak*-topology. Then E is a locally convex Hausdorff topological vector space and, by the Banach–Alaoglu Theorem (see e.g. [Con90], Chapter 5, Section 3), any closed ball in E is weak*-compact. The set of all means on $\mathcal{CB}(K)$ is a convex subset of the weak*-compact unit ball of E and we show that it is also weak*-closed. Let N denote a linear functional from the weak*-closure of the set of all means. Suppose that $Im\, N(f) \neq 0$ for some real valued f in $\mathcal{CB}(K)$. Then the set

$$\{\psi : |\psi(f) - N(f)| < |Im\, N(f)|\}$$

is a weak*-neighbourhood of N, hence it contains a mean φ, which is a contradiction. Hence $N(f)$ is real for real valued functions f. Suppose now that $N(f) < 0$ for some real valued and nonnegative f in $\mathcal{CB}(K)$. Then the set

$$\{\psi : |\psi(f) - N(f)| < -N(f)\}$$

is a weak*-neighbourhood of N, hence it contains a mean φ, but $\varphi(f)$ is nonnegative, which is a contradiction. Hence $N(f)$ is nonnegative for nonnegative real valued functions f. Finally, for any $\varepsilon > 0$ the set

$$\{\psi : |\psi(1) - N(1)| < \varepsilon\}$$

is a weak*-neighbourhood of N, hence it contains a mean φ, but $\varphi(1) = 1$, hence $N(1) = 1$. By Theorem 1.3 it follows that N is a mean and the set of all means is a weak*-compact set.

For any y in K and for any linear functional φ in E we define the functional $T_y\varphi$ on $\mathcal{CB}(K)$ by the formula

$$T_y\varphi(f) = \varphi(\tau_y f).$$

It is easy to see that $T_y\varphi$ is in E and T_y is a weak*-continuous linear endomorphism of E. It is also clear that T_y maps the set of all means into itself and the operators T_y and T_z commute for any y, z in K. By the Markov–Kakutani Theorem all these operators have a common fixed point M in the set of all means. It is obvious that M is an invariant mean on K, hence K is. $\qquad\square$

The interested reader will find further results and references concerning invariant means, invariant measures and amenability on hypergroups in [LS11], [Azi10], [Pav07a], [KG06], [Raj02], [Geb98].

1.4   Haar measure on hypergroups

The existence of right and left translations in hypergroups makes it possible to introduce also right and left invariant measures. Let K be a hypergroup. A measure μ in $\mathcal{M}_+(K)$ is called *right invariant* if

$$\int_K \tau_y f \, d\mu = \int_K f \, d\mu$$

holds for each y in K and for any f in $\mathcal{C}_c(K)$. We define *left invariant* measures analogously, using left translations and we call a measure in $\mathcal{M}_+(K)$ *invariant* if it is right and left invariant. Right invariant, left invariant and invariant measures are also called *right Haar measure, left Haar measure* and *Haar measure*, respectively. The question about the existence and uniqueness of right or left Haar measure on hypergroups is a nontrivial problem, however, it has been answered in the positive in the most important cases. First we consider the existence of Haar measure on commutative hypergroups. We follow the ideas of [Izz92] (see also [BH95]).

We need the following theorem.

Theorem 1.6. *Let K be a commutative hypergroup and let U be a symmetric neighbourhood of the identity. Then there exists a subset S in K such that for any a in K the set $a * U * U$ contains at least one element of S and the set $a * U$ contains at most one element of S.*

Proof. Let $\mathcal{K}$ denote the collection of all subsets T of K with the property that

$$q \notin p * U * U$$

for each p, q in T, whenever $p \neq q$. From this condition it follows easily that $p \notin q * U * U$. Indeed, if $p \in q * U * U$, then

$$\{p\} \cap q * U * U \neq \emptyset$$

and we can apply Theorem 1.1 with the choice $A = U * U$, $B = \{p\}$ and $C = \{q\}$. Then, by the symmetric property of U, we have

$$\{p\} \cap q * U * U \neq \emptyset \, ,$$

that is

$$p \in q * U * U \, .$$

Obviously, each subset of K consisting of at most one element belongs to $\mathcal{K}$ and it is easy to see that $\mathcal{K}$ satisfies the conditions of Zorn's Lemma, hence it has a maximal element S. Let a be an arbitrary element of K and suppose that $a * U * U$ contains no element of S that is

$$S \cap a * U * U = \emptyset \, .$$

We apply Theorem 1.1 with the choice $A = U * U$, $B = \{a\}$ and $C = S$, then, by the symmetry of U, it follows

$$\{a\} \cap U * U * S = \emptyset \, ,$$

that is, a does not belong to the set $U * U * S$, hence a does not belong to the set $s * U * U$. By the above considerations this implies that s does not belong to the set $a * U * U$, hence the set $S \cup \{a\}$ belongs to $\mathcal{K}$ and this is a contradiction, as it strictly contains S. This means that for any a in K the set $a * U * U$ contains at least one element of S.

On the other hand, suppose that there are elements $s_1 \neq s_2$ in S such that

$$s_1 \in a * U \, , \qquad \text{and} \qquad s_2 \in a * U \, .$$

Then $\{s_1\} \cap (a * U) \neq \emptyset$ and we apply Theorem 1.1 again, with the choice $A = A^{\vee} = U$, $B = \{s_1\}$ and $C = \{a\}$. It follows

$$(U * s_1) \cap \{a\} \neq \emptyset ,$$

hence $a \in s_1 * U$ and $a^\vee \in s_1^\vee * U$. Similarly, we have $a \in s_2 * U$. From this we infer that

$$supp\,(a * a^\vee) \subseteq s_1^\vee * s_2 * U * U .$$

By the definition of a hypergroup we have that e is in $s_1^\vee * s_2 * U * U$ that is

$$\{e\} \cap (s_1^\vee * s_2 * U * U) \neq \emptyset .$$

We apply Theorem 1.1 with the choice $A = \{s_1\}$, $B = \{e\}$ and $C = s_2 * U * U$ to obtain

$$(\{s_1\} * \{e\}) \cap (s_2 * U * U) \neq \emptyset ,$$

that is

$$\{s_1\} \cap (s_2 * U * U) \neq \emptyset ,$$

which implies that s_1 belongs to $s_2 * U * U$. This contradicts the definition of S. Hence the set $a * U$ contains at most one element of S and the theorem is proved. □

Theorem 1.7. *Every commutative hypergroup admits a Haar measure.*

Proof. The proof is based on the Markov–Kakutani Theorem 1.4. Let the commutative hypergroup K be given. Then K is a locally compact Hausdorff space. The space $C_c(K)^*$ denotes the space of all linear functionals of $C_c(K)$ equipped with the weak*-topology. For each a in K we denote the mapping $T_a : C_c(K)^* \to C_c(K)^*$ by the formula

$$T_a \, \Lambda(f) = \Lambda(\tau_a f)$$

for every f in $C_c(K)$. Then each T_a is a continuous linear operator on the Hausdorff topological vector space $C_c(K)^*$. Clearly, the operators T_a commute. In order to prove the theorem it is enough to show that there exists a nonzero positive linear functional on $C_c(K)$, which is fixed by each T_a.

As soon as we can define a nonempty compact convex subset C in $C_c(K)^*$, which is mapped into itself by all operators T_a, the Markov–Kakutani Theorem 1.4 will imply our statement. First we fix a symmetric neighbourhood U of the identity in K with compact closure. Let C be the set of all positive linear functionals Λ with the following two properties:

(1) $\Lambda(f) \leq 1$ whenever f is in $C_c(K)$ and $0 \leq f \leq 1$, further the support of f is contained in $a * U$ for some a in K;

(2) $\Lambda(f) \geq 1$ whenever f is in $C_c(K)$ and $0 \leq f$, further $f = 1$ on $a * U * U$ for some a in K.

Clearly, f is convex and closed in $C_c(K)^*$. By a partition of unity argument every nonnegative function in $C_c(K)$ can be written as a finite sum of nonnegative continuous functions each of which has support in $a * U$ for some a in K. Hence the condition (1) in the definition of C implies that for every function f in $C_c(K)$ the set $\{\Lambda(f) : \Lambda \in C\}$ is bounded. Therefore, by Lemma 2 in [Izz92], the set C is compact. Clearly, all operators T_a map C into itself. We have to show that C is nonempty. Indeed, if S is the set in Theorem 1.6 chosen to U, then the linear functional $f \mapsto \sum_{s \in S} f(s)$ belongs to K.

By the Markov–Kakutani Theorem 1.4 we infer that all operators T_a with a in K have a common fixed point, which is a Haar integral and the theorem is proved. $\square$

Another important special case is represented by the discrete hypergroups, where the existence and uniqueness of one-sided Haar measure can be proved relatively easily. The following proof is taken from [BH95].

Theorem 1.8. *Every discrete hypergroup admits a right Haar measure ω_K, which is unique up to a positive constant. It can be normalized to satisfy $\omega_K(\{e\}) = 1$ and in this case*

$$\omega_K(\{x\}) = \big((\delta_{x^\vee} * \delta_x)(\{e\})\big)^{-1} \tag{1.4}$$

for each x in K.

Proof.   For x, y, z in K define $[x * y, z] = (\delta_x * \delta_y)(\{z\})$. Note that $[x * y, e] > 0$ if and only if $x = y^\vee$. For each x in K let

$$[x] = \frac{1}{[x^\vee * x, e]} \cdot$$

If x, y, z are in K, then we have

$$\big(\delta_x * (\delta_y * \delta_z)\big)(\{e\}) = \big((\delta_x * \delta_y) * \delta_z\big)(\{e\}),$$

$$\sum_{t \in K} [x * t, e]\,[y * z, t] = \sum_{t \in K} [x * y, t]\,[t * z, e],$$

$$[x * x^\vee, e]\,[y * z, x^\vee] = [x * y, z^\vee]\,[z^\vee * z, e],$$

$$[z]\,[y * z, x^\vee] = [x^\vee]\,[x * y, z^\vee].$$

Let the measure ω_K be defined by

$$\omega_K = \sum_{x \in K} [x]\,\delta_x\,.$$

If x, y are in K, then

$$(\delta_y * \omega_K)(\{x^\vee\}) = \sum_{z \in K} [z](\delta_y * \delta_z)(\{x^\vee\})$$

$$= \sum_{z \in K} [z]\,[\delta_y * \delta_z, \{x^\vee\}] = \sum_{z \in K} [x^\vee]\,[\delta_x * \delta_y, \{z^\vee\}] = [x^\vee] = \omega_K(\{x^\vee\})\,.$$

$\square$

Obviously, one can prove similarly that also a unique (apart from constant multiple) left Haar measure exists on discrete hypergroups. However, we note that in contrast to the group case it is unknown whether all discrete hypergroups admit a Haar measure. Nevertheless, all compact hypergroups admit a Haar measure, as it is proved in the following theorem. For this we need another fixed point theorem of Kakutani and some preliminary lemmas.

Theorem 1.9. *(Kakutani) Let K be a nonempty compact convex set in a locally convex topological vector space X, let G be an equicontinuous group of linear mappings of X into X, further let $\Lambda(K) \subseteq K$ for every Λ in G. Then G has a common fixed point in K, that is, there exists a p in K such that $\Lambda p = p$ for every Λ in G.*

The proof of this theorem can be found in [Rud73], p. 120.

Theorem 1.10. *Let K be a compact Hausdorff space, f a continuous complex valued function on K, further let $H_\tau(f)$ denote the convex hull of the set of all left translates of f. Then $H_\tau(f)$ is a totally bounded subset of $\mathcal{C}(K)$.*

Proof. We get the proof and the statement along the lines of [Rud73] followed by Ascoli's Theorem (see [Rud73], pp. 122 and 369). □

Theorem 1.11. *Every compact hypergroup admits a Haar measure which is unique up to a positive constant.*

Proof. The right translation operators τ_s satisfy $\tau_s \tau_t = \tau_{ts}$ as

$$(\tau_s \tau_t f)(x) = (\tau_t f)(s * x) = f(t * s * x) = (\tau_{ts} f)(x).$$

As each τ_s is an isometry of $\mathcal{C}(K)$ onto itself, hence $\{\tau_s : s \in G\}$ is an equicontinuous group of linear operators on $\mathcal{C}(K)$. If f is in $\mathcal{C}(K)$, then let K_f be the closure of $(H_\tau(f)$. By the previous theorem K_f is compact. It is obvious that $\tau_s(K_f) = K_f$ for each s in G. By the Kakutani fixed-point Theorem 1.9 the set K_f contains a function Φ such that $\tau_s \Phi = \Phi$ for each s in K. In particular, $\Phi(s) = \Phi(e)$, (e is the identity), so that Φ is constant. By the definition of K_f this constant can be uniformly approximated by functions in $H_\tau(f)$.

We have proved that for each f in $\mathcal{C}(K)$ there is a constant c, which can be uniformly approximated on K by convex combinations of left translates of f. Similarly, there is a constant d with a similar property concerning the right translates of f. We show that $c = d$.

Let $\varepsilon > 0$ be given. Then there exist finite sets $\{a_i\}$ and $\{b_j\}$ in K and positive nombers α_i, β_j with $\sum_i \alpha_i = \sum_j \beta_j = 1$ such that

$$\left| c - \sum_i \alpha_i f(a_i * x) \right| < \varepsilon \tag{1.5}$$

and

$$\left| d - \sum_j \beta_j f(x * b_j) \right| < \varepsilon \tag{1.6}$$

holds for each x in K. Put $x = b_j$ in (1.5), multiply (1.5) by β_j and add the equations with respect to j. Then we have

$$\left| c - \sum_{i,j} \alpha_i \beta_j f(a_i * b_j) \right| < \varepsilon. \tag{1.7}$$

Similarly, put $x = a_i$ in (1.6), multiply (1.6) by α_i and add the equations with respect to i. Then we obtain

$$\left| d - \sum_{i,j} \alpha_i \beta_j f(a_i * b_j) \right| < \varepsilon. \tag{1.8}$$

From (1.7) and (1.8) it follows $c = d$.

Hence we have that to each f in $\mathcal{C}(K)$ there corresponds a unique number $M(f)$, which can be uniformly approximated by convex combinations of left translates of f and also it can be uniformly approximated by convex combinations of right translates of f. Obviously, the following properties hold for each f in $\mathcal{C}(K)$:

(1) $M(f) \geq 0$ for $f \geq 0$,

(2) $M(1) = 1$,

(3) $M(\alpha \cdot f) = \alpha \cdot M(f)$ for each complex number α,

(4) $M(_y\tau\, f) = M(\tau_y\, f) = M(f)$ for each y in K.

We prove that

$$M(f + g) = M(f) + M(g) \tag{1.9}$$

holds for all f, g in $\mathcal{C}(K)$. Let $\varepsilon > 0$ be arbitrary. Then for each x in K

$$\left| M(f) - \sum_i \alpha_i f(a_i * x) \right| < \varepsilon \qquad (1.10)$$

holds for some finite set $\{a_i\}$ in K and for some positive numbers α_i with $\sum_i \alpha_i = 1$. We define

$$h(x) = \sum_i \alpha_i g(a_i * x) \qquad (1.11)$$

for each x in K. Then h belongs to K_g, hence $K_h \subseteq K_g$ and since both sets contain a unique constant function, we have $M(h) = M(g)$. Hence there is a finite set $\{b_j\}$ in K and there are positive numbers β_j with $\sum_j \beta_j = 1$ such that for all x in K we have

$$\left| M(g) - \sum_j \beta_j h(b_j * x) \right| < \varepsilon . \qquad (1.12)$$

By the definition (1.11) this gives

$$\left| M(g) - \sum_{i,j} \alpha_i \beta_j g(a_i * b_j * x) \right| < \varepsilon \qquad (1.13)$$

for each x in K. By (1.10) it follows

$$\left| M(f) - \sum_{i,j} \alpha_i \beta_j f(a_i * b_j * x) \right| < \varepsilon \qquad (1.14)$$

for each x in K. Hence, by (1.13) and (1.14), we have

$$\left| M(f) + M(g) - \sum_{i,j} \alpha_i \beta_j (f + g)(a_i * b_j * x) \right| < 2\varepsilon \qquad (1.15)$$

for each x in K. As $\sum \alpha_i \beta_j = 1$, (1.15) implies (1.9). This means that M is a Haar integral on $\mathcal{C}(K)$, which – by the Riesz Representation Theorem – arises from a Haar measure. The proof is complete. $\qquad \square$

As we have seen in the case of discrete hypergroups the right Haar measure ω_K can be normalized to satisfy $\omega_K(\{e\}) = 1$ and in the case of compact hypergroups the usual normalization is $\omega_K(K) = 1$. In case of finite hypergroups one prefers the discrete normalization.

As a simple illustration we compute here the Haar measure ω_D on the hypergroup $D(\theta)$. By assumption

$$f(0)\omega_D(\{0\}) + f(1)\omega_D(\{1\}) = f(0*1)\omega_D(\{0\}) + f(1*1)\omega_D(\{1\})$$

holds for any function $f : D(\theta) \to \mathbb{C}$. Using the definition of convolution in $D(\theta)$ we have

$$f(0)\omega_D(\{0\})+f(1)\omega_D(\{1\}) = f(1)\omega_D(\{0\})+\big(\theta f(0)+(1-\theta)f(1)\big)\omega_D(\{1\}),$$

or, equivalently

$$\Big(f(0) - f(1)\Big)\Big(\omega_D(\{0\}) - \theta\omega_D(\{1\})\Big) = 0.$$

This equation holds for any choice of $f(0)$ and $f(1)$, which implies that

$$\omega_D(\{0\}) - \theta\omega_D(\{1\}) = 0,$$

hence, by normalization

$$\omega_D(\{0\}) = 1 \quad \text{and} \quad \omega_D(\{1\}) = \frac{1}{\theta}.$$

The existence of Haar measure on a wide class of hypergroups makes it possible to adopt integration theory and harmonic analysis on this class similarly, as in the case of locally compact topological groups. In the presence of Haar measure the induced integral is called *Haar integral* and the corresponding Lebesgue spaces will be denoted in the usual manner $L^p(\omega_K)$ for $1 \le p \le +\infty$.

The following simple result is very easy to prove.

Theorem 1.12. *If A is a compact subset of the hypergroup K with Haar measure ω, then $\omega(A) \le \omega(x*A)$ holds for each x in K.*

Sometimes we need a Steinhaus-type theorem.

Theorem 1.13. *Let K be a hypergroup with Haar measure ω and let A be a measurable set of positive finite measure. Then $A * A^{\vee}$ contains a neighbourhoood of e.*

Proof. Since Haar measure is regular, we may assume that A is compact. We choose an open set $U \supset A$ such that $\omega(U) < 2\omega(A)$. Now choose an open neighbourhood V of e satisfying $V * A \subset U$. Then $(v*A) \cap A \neq \emptyset$ for all v in V, otherwise for some v in V we have $\omega(U) \geq \omega(v*A)+\omega(A) \geq 2\omega(A)$, where the last inequality follows from the previous theorem and this would contradict the choice of U. Hence, for all v in V we have that v belongs to $A * A^{\vee}$, that is, $V \subseteq A * A^{\vee}$.   □

1.5   Exponential functions on hypergroups

Exponential functions play a fundamental role in the theory of functional equations on group-like algebraic structures, where translation operators appear. Namely, they serve as homomorphisms of the underlying structure into the multiplicative group of nonzero complex numbers. Harmonic analysis depends on special exponential functions, called characters. The presence of translation operators on hypergroups makes it possible to introduce the concept of exponential functions and that of characters.

Let K be a commutative hypergroup with convolution $*$, involution $^{\vee}$ and identity e. For any y in K let τ_y denote the translation operator corresponding to the element y in K on the space of all complex valued functions on K, which are integrable with respect to $\delta_x * \delta_y$ for each x, y in K. In particular, any continuous complex valued function belongs to this class. We call the continuous complex valued function $m : K \to \mathbb{C}$ on K an *exponential function* or simply an *exponential* if it is not identically zero and

$$\tau_y m(x) = m(x)m(y)$$

holds for all x, y in K. In other words, m satisfies the functional equation

$$\int_K m(t)\, d(\delta_x * \delta_y)(t) = m(x)m(y)$$

for all x, y in K. Real-valued exponentials are called *real exponentials*.

As we indicated above, we can write the above equation in the more suggestive form

$$m(x * y) = m(x)m(y).$$ (1.16)

Putting $y = e$ into this equation it follows that $m(e) = 1$ holds for any exponential. If m also satisfies $m(x^{\vee}) = \overline{m(x)}$ for every x in K, then we call m a *semi-character*. Any bounded semi-character is called a *character*. Hence on Hermitian hypergroups semi-characters are exactly the real exponentials and on compact hypergroups any semi-character is a character. We remark that in the terminology of [BH95] exponentials are not necessarily continuous. We underline the inconvenient fact that, in contrast to the case of groups, exponential functions can take the zero value and the product of two exponentials is not necessarily an exponential.

From the above definition it is clear that exponential functions on hypergroups are solutions of special integral equations with respect to measures related to the hypergroup structure. It is also clear that in the group-case the problems concerning exponentials reduce to the classical problems about exponential functions. However, in the general case we might expect very interesting, sometimes surprising situations. Here we give a simple illustration of this fact by describing all exponential functions on the hypergroup $D(\theta)$ with $0 < \theta \leq 1$.

Suppose that $m : D(\theta) \to \mathbb{C}$ is an exponential that is

$$\int_{D(\theta)} m(t) \, d(\delta_x * \delta_y)(t) = m(x)m(y)$$

holds for each x, y in $D(\theta)$. According to the definition of convolution in $D(\theta)$ the only nontrivial consequence of this equation we obtain in the case $x = y = 1$:

$$\theta m(0) + (1 - \theta)m(1) = m(1)m(1).$$

Using $m(0) = 1$ and solving the quadratic equation for $m(1)$ we have the two possibilities: $m(1) = 1$ or $m(1) = -\theta$. The first case gives the trivial exponential which is identically 1 and the second case is the nontrivial one: $m(0) = 1$ and $m(1) = -\theta$.

The set $\widetilde{K}$ of all exponentials on the hypergroup K will be given the topology of uniform convergence on compact sets and it will be called the *generalized dual of K*. The subspace $\widehat{K}$ of all characters of K is called the *dual of K*. Given a bounded measure μ in $\mathcal{M}_b(K)$ the function $\widehat{\mu} : \widehat{K} \to \mathbb{C}$ defined by

$$\widehat{\mu}(\chi) = \int_K \chi^\vee \, d\mu$$

for any character χ is called the *Fourier transform* of the measure μ. If μ is a probability measure, then $\widehat{\mu}$ is called the *characteristic function* of μ. For any compactly supported measure μ in $\mathcal{M}_c(K)$ this can be extended to $\widetilde{K}$ by the same formula and the extension is called the *Fourier–Laplace transform* of μ. We shall use the same notation for this extension. Finally, given any function f in $L^1(\omega_K)$ the function $\widehat{f} : \widehat{K} \to \mathbb{C}$ defined by

$$\widehat{f}(\chi) = \int_K f\chi^\vee \, d\omega_K$$

for any character χ is called the *Fourier transform* of the function f. For more about Fourier transforms and Fourier–Laplace transforms of measures and functions the reader should refer to [BH95]. Here we quote the convolution formula for compactly supported measures μ, ν and for integrable functions f, g:

$$(\mu * \nu)\widehat{} = \widehat{\mu} \cdot \widehat{\nu}$$

and

$$(f * g)\widehat{} = \widehat{f} \cdot \widehat{g}.$$

1.6   Exponential families on hypergroups

Let K be a commutative hypergroup and n a positive integer. Suppose that $\Phi : K \times \mathbb{C}^n \to \mathbb{C}$ is a function with the following properties:

(1) The function $x \mapsto \Phi(x, \lambda)$ is an exponential on K for each λ in $\mathbb{C}^n$.

(2) The function $\lambda \mapsto \Phi(x, \lambda)$ is entire for each x in K.

(3) For each exponential m on K there exists a unique λ in $\mathbb{C}^n$ such that $m(x) = \Phi(x, \lambda)$ for all x in K.

In this case Φ is called an *exponential family* on the hypergroup K. This means that any exponential family Φ on K satisfies

$$\Phi(x * y, \lambda) = \Phi(x, \lambda) \, \Phi(y, \lambda) \qquad (1.17)$$

for all x, y in K and λ in $\mathbb{C}^n$. By the third property of Φ there exists a unique λ_0, such that $\Phi(x, \lambda_0) = 1$ for each x in K. Obviously, we always may suppose that $\lambda = 0$ and we shall do this in the sequel.

1.7 Additive and multi-additive functions on hypergroups

Besides exponentials another important class of functions on commutative groups is formed by additive functions, which are homomorphisms into the additive group of complex numbers. The theory of convolution-type functional equations on topological Abelian groups is based on exponential and additive functions (see [Szé91], [Szé06a]). The existence of translation operators on hypergroups makes it possible to define this important function class on hypergroups.

The continuous complex valued function $a : K \to \mathbb{C}$ on the hypergroup K is called an *additive function*, if

$$\tau_y a(x) = a(x) + a(y)$$

holds for all x, y in K. In more details this means that a satisfies

$$\int_K a(t) \, d(\delta_x * \delta_y)(t) = a(x) + a(y)$$

for each x, y in K. Using the "group-like" notation this has the form

$$a(x * y) = a(x) + a(y). \qquad (1.18)$$

Substituting $y = e$ we have $a(e) = 0$ for any additive function. It is clear that any linear combination of additive functions is additive again. However, in contrast to the group-case, some simple relations between exponential and additive functions on groups are no longer valid: logarithms of positive exponentials are not necessarily additive and $\exp a$ is not necessarily an exponential for an additive function a.

As an illustration we determine all additive functions on $D(\theta)$. Supposing that $a : D(\theta) \to \mathbb{C}$ is additive it satisfies

$$\int_{D(\theta)} a(t) \, d(\delta_x * \delta_y)(t) = a(x) + a(y)$$

for each x, y in $D(\theta)$. Here the only nontrivial consequence is

$$\theta a(0) + (1 - \theta)a(1) = a(1) + a(1) = 2a(1) \,,$$

and by $a(0) = 0$ this implies $a(1) = 0$, hence the only additive function on $D(\theta)$ is identically zero. It is easy to see that the same holds for any compact hypergroup.

Suppose that K is a commutative hypergroup with exponential family $\Phi : K \times \mathbb{C}^n \to \mathbb{C}$. If we let $\lambda = (\lambda_1, \lambda_2, \ldots, \lambda_n)$ for each λ in $\mathbb{C}^n$, then we can realize Φ as an $n + 1$-place function defined on $K \times \mathbb{C} \times \mathbb{C} \times \cdots \times \mathbb{C}$, where n copies of $\mathbb{C}$ appear. Let $a_k : K \to \mathbb{C}$ be defined for $k = 1, 2, \ldots, n$ as

$$a_k(x) = \partial_{k+1}\Phi(x, 0) \,,$$

where ∂_{k+1} means differentiation with respect to the k-th coordinate of the second variable of Φ. It is easy to see that a_k is an additive function on K for each $k = 1, 2, \ldots, n$. Indeed, if we differentiate both sides of the equation (1.17) with respect to the $k + 1$-th variable and substitute $\lambda = 0 = (0, 0, \ldots, 0)$ we get

$$a_k(x * y) = \partial_{k+1}\Phi(x, 0) \cdot \Phi(y, 0) + \Phi(x, 0) \cdot \partial_{k+1}\Phi(y, 0) = a_k(x) + a_k(y)$$

for each x, y in K, as – by the assumptions on Φ – $\Phi(x, 0) = 1$ for each x in K. This means that any exponential family Φ on the commutative hypergroup K generates a family of additive functions on K, namely the linear space of all linear combinations of the form

$$x \mapsto \sum_{k=1}^{n} c_k \, \partial_{k+1} \Phi(x, 0)$$

with arbitrary complex numbers c_k. We shall see that in some cases this linear space contains all additive functions on K, but this is not necessarily the case in general. Suppose, for instance that K is a commutative hypergroup with the exponential family $\Phi : K \times \mathbb{C} \to \mathbb{C}$ and we assume that the functions $x \mapsto \partial_2 \Phi(x, 0), x \mapsto \partial_2^2 \Phi(x, 0), \ldots, x \mapsto \partial_2^k \Phi(x, 0)$ are identically zero on K for some positive integer k. Then it is easy to check that $x \mapsto \partial_2^{k+1} \Phi(x, 0)$ is an additive function on K. In this case, if this function is not identically zero, then it is an additive function, which is obviously not contained in the linear space mentioned above. We shall see an example for this situation in the sequel.

The definition of multi-additive functions is straightforward. For any positive integer n we say that the continuous function $A : K^n \to \mathbb{C}$ is n-*additive*, if it is additive in each variable while the others are fixed. Hence 1-additive functions are simply the additive ones and 2-additive functions are called *bi-additive*. Extending our terminology to the case $n = 0$ we may consider the zero function 0-additive.

Another important class is formed by quadratic functions. The continuous function $q : K \to \mathbb{C}$ is called *quadratic* if it satisfies

$$q(x * y) + q(x * y^{\vee}) = 2q(x) + 2q(y)$$

for each x, y in K. On Hermitian hypergroups quadratic functions are exactly the additive ones. However, the square of an additive function is not necessarily quadratic in general. The characterization of quadratic functions on different commutative hypergroups is an interesting problem.

1.8   Moment functions on hypergroups

Let K be a hypergroup and N a positive integer. Moments of probability measures on a hypergroup can be introduced in terms of moment functions. The notion of moment functions has been formalized in [Zeu92] (see also [BH95]). Concerning methods of finding moment functions on hypergroups and other results on moment function sequences see [Gal98], [OS05], [OS04], [OS08], [Zeu92], [Vaj10b], [Szé06c], [Gal97].

For any nonnegative integer N the continuous function $\varphi : K \to \mathbb{C}$ is called a *moment function of order* N, if there exist complex valued continuous functions $\varphi_k : K \to \mathbb{C}$ for $k = 0, 1, \ldots, N$ such that $\varphi_0 = 1$, $\varphi_N = \varphi$ and

$$\varphi_k(x * y) = \sum_{j=0}^{k} \binom{k}{j} \varphi_j(x) \varphi_{k-j}(y) \qquad (1.19)$$

holds for $k = 0, 1, \ldots, N$ and for all x, y in K. In this case we say that the functions φ_k ($k = 0, 1, \ldots, N$) form a *moment function sequence of order* N. In particular, moment functions of order 1 are exactly the additive functions. In the applications *associated pairs* of real valued first and second order moment functions (φ_1, φ_2) play a distinguished role, which means that they are subjected to the additional condition

$$\varphi_1(x)^2 \leq \varphi_2(x) \qquad (1.20)$$

for each x in K. In particular, φ_2 is nonnegative. In this case the system of equations (1.19) reduces to the pair of functional equations

$$\varphi_1(x * y) = \varphi_1(x) + \varphi_1(y) \qquad (1.21)$$

and

$$\varphi_2(x * y) = \varphi_2(x) + 2\varphi_1(x)\varphi_1(y) + \varphi_2(y) \qquad (1.22)$$

for all x, y in K. If φ_1 is identically zero, then φ_2 is additive and we call this pair *trivial*.

With respect to a nontrivial associated pair of moment functions and for any probability measure μ in $\mathcal{M}_1(K)$ with

$$\int_K \varphi_2 \, d\mu < +\infty$$

one defines its *generalized expectation* by

$$\mathbb{E}(\mu) = \int_K \varphi_1 \, d\mu \tag{1.23}$$

and *generalized variance* by

$$\mathbb{V}(\mu) = \int_K \varphi_2 \, d\mu - \left[\int_K \varphi_1 \, d\mu \right]^2 . \tag{1.24}$$

By equations (1.21) and (1.22) the operators $\mathbb{E}$ and $\mathbb{V}$ have the additive property: for any probability measures μ, ν in $\mathcal{M}_1(K)$ satisfying

$$\int_K \varphi_2 \, d\mu < +\infty \qquad \text{and} \qquad \int_K \varphi_2 \, d\nu < +\infty$$

we have

$$\mathbb{E}(\mu * \nu) = \mathbb{E}(\mu) + \mathbb{E}(\nu) \qquad \text{and} \qquad \mathbb{V}(\mu * \nu) = \mathbb{V}(\mu) + \mathbb{V}(\nu) .$$

We can generalize the above concepts by omitting the hypothesis $\varphi_0 = 1$. In this case φ_0 is an exponential function and we say that φ_0 *generates the generalized moment function sequence of order N*, further φ_k is a *generalized moment function of order k with respect to φ_0* $(k = 0, 1, \ldots, N)$. In other words, we say that the functions $\varphi_k : K \to \mathbb{C}$ $(k = 0, 1, \ldots, N)$ form a *sequence of generalized moment functions* (of order N). For instance, generalized moment functions of order 1 with respect to the exponential φ_0 are exactly the solutions of the *sine functional equation*

$$\varphi_1(x * y) = \varphi_0(x)\varphi_1(y) + \varphi_0(y)\varphi_1(x)$$

for each x, y in K. If $\varphi_0, \varphi_1, \varphi_2$ form a generalized moment function sequence of order 2 generated by the exponential φ_0, then (φ_1, φ_2) is called *a pair of generalized moment functions generated by* φ_0. In case of generalized moment functions the definition of associated pair should be modified as follows: Let φ_0 be an exponential on the hypergroup K and let $\varphi_0, \varphi_1, \varphi_2$ form a generalized moment sequence of order 2 generated by the exponential φ_0. If all these functions are real valued, φ_2 is nonnegative, further we have

$$\varphi_1(x)^2 \leq \varphi_0(x) \cdot \varphi_2(x) \qquad (1.25)$$

for each x in K, then (φ_1, φ_2) is called a *generalized associated pair* generated by the exponential φ_0.

Let $(K, *)$ be any commutative hypergroup and let μ be a probability measure in $\mathcal{M}_1(K)$. We say that the exponential function m on K is *not in the spectrum of* μ, if m is integrable with respect to μ and $\int_K m \, d\mu \neq 0$. In this sense the *spectrum of* μ consists of all exponentials m on K, which are either non-integrable with respect to μ, or satisfy $\int_K m \, d\mu = 0$. In particular, if K is Hermitian and μ is compactly supported, then its spectrum is the set of all zeros of its Fourier–Laplace transform.

Now we fix a real valued exponential $\varphi_0 : K \to \mathbb{R}$ on K and a generalized associated pair generated by the exponential φ_0. If the exponential φ_0 does not belong to the spectrum of the probability measure μ, further φ_2 is integrable with respect to μ, then we define

$$\mathbb{E}_{\varphi_0}(\mu) = \frac{\int_K \varphi_1 \, d\mu}{\int_K \varphi_0 \, d\mu}$$

and

$$\mathbb{V}_{\varphi_0}(\mu) = \frac{\int_K \varphi_2 \, d\mu}{\int_K \varphi_0 \, d\mu} - \left(\frac{\int_K \varphi_1 \, d\mu}{\int_K \varphi_0 \, d\mu} \right)^2$$

as the *generalized expectation* and *generalized variance of* μ *with respect to* φ_0. Obviously we have

$$\mathbb{V}_{\varphi_0}(\mu) = \frac{\int_K \varphi_2 \, d\mu}{\int_K \varphi_0 \, d\mu} - \mathbb{E}_{\varphi_0}(\mu)^2 \, .$$

These quantities have the following remarkable properties.

Theorem 1.14. *Let $(K, *)$ be a commutative hypergroup and let $\varphi_0, \varphi_1, \varphi_2$ be a sequence of continuous real valued generalized moment functions with nonnegative φ_2 satisfying*

$$\varphi_1(x)^2 \leq \varphi_0(x)\, \varphi_2(x)$$

for all x in K. If μ, ν are probability measures in $\mathcal{M}_1(K)$ such that φ_0 is not in the spectrum of μ and ν, further φ_2 is integrable with respect to μ and ν, then the generalized expectation and generalized variance of μ, ν and $\mu * \nu$ with respect to φ_0 exist and we have

$$\mathbb{E}_{\varphi_0}(\mu * \nu) = \mathbb{E}_{\varphi_0}(\mu) + \mathbb{E}_{\varphi_0}(\nu), \qquad \mathbb{V}_{\varphi_0}(\mu * \nu) = \mathbb{V}_{\varphi_0}(\mu) + \mathbb{V}_{\varphi_0}(\nu).$$

Proof. By the Fubini Theorem and the convolution formula

$$(\mu * \nu)\hat{} = \hat{\mu} \cdot \hat{\nu}$$

it follows that φ_0 is not in the spectrum of $\mu * \nu$. The existence of the generalized expectation and generalized variance of μ, ν and $\mu * \nu$ with respect to φ_0 is a consequence of the Cauchy–Schwartz Inequality and of the condition (1.25). On the other hand,

$$\mathbb{E}_{\varphi_0}(\mu * \nu) = \frac{\int_K \varphi_1 \, d(\mu * \nu)}{\int_K \varphi_0 \, d(\mu * \nu)} = \frac{\int_K \int_K \varphi_1(x * y) \, d\mu(x) \, d\nu(y)}{\int_K \int_K \varphi_0(x * y) \, d\mu(x) \, d\nu(y)}$$

$$= \frac{\int_K \int_K \varphi_1(x)\varphi_0(y) \, d\mu(x) \, d\nu(y) + \int_K \int_K \varphi_1(y)\varphi_0(x) \, d\mu(x) \, d\nu(y)}{\int_K \int_K \varphi_0(x)\varphi_0(y) \, d\mu(x) \, d\nu(y)}$$

$$= \frac{\int_K \varphi_1(x) \, d\mu(x) \int_K \varphi_0(y) \, d\nu(y) + \int_K \varphi_1(y) \, d\nu(y) \int_K \varphi_0(x) \, d\mu(x)}{\int_K \varphi_0(x) \, d\mu(x) \int_K \varphi_0(y) \, d\nu(y)}$$

$$= \frac{\int_K \varphi_1(x) \, d\mu(x)}{\int_K \varphi_0(x) \, d\mu(x)} + \frac{\int_K \varphi_1(y) \, d\nu(y)}{\int_K \varphi_0(y) \, d\nu(y)} = \mathbb{E}_{\varphi_0}(\mu) + \mathbb{E}_{\varphi_0}(\nu) \, .$$

Similarly, we can compute as follows:

$$\mathbb{V}_{\varphi_0}(\mu * \nu) = \frac{\int_K \varphi_2 \, d(\mu * \nu)}{\int_K \varphi_0 \, d(\mu * \nu)} - \left(\frac{\int_K \varphi_1 \, d(\mu * \nu)}{\int_K \varphi_0 \, d(\mu * \nu)} \right)^2$$

$$= \frac{\int_K \int_K \varphi_2(x * y) \, d\mu(x) \, d\nu(y)}{\int_K \int_K \varphi_0(x * y) \, d\mu(x) \, d\nu(y)} - \left(\frac{\int_K \int_K \varphi_1(x * y) \, d\mu(x) \, d\nu(y)}{\int_K \int_K \varphi_0(x * y) \, d\mu(x) \, d\nu(y)} \right)^2$$

$$= \frac{\int_K \int_K \varphi_2(x)\varphi_0(y) \, d\mu(x) \, d\nu(y)}{\int_K \int_K \varphi_0(x)\varphi_0(y) \, d\mu(x) \, d\nu(y)} + 2 \frac{\int_K \int_K \varphi_1(x)\varphi_1(y) \, d\mu(x) \, d\nu(y)}{\int_K \int_K \varphi_0(x)\varphi_0(y) \, d\mu(x) \, d\nu(y)}$$

$$+ \frac{\int_K \int_K \varphi_2(y)\varphi_0(x) \, d\mu(x) \, d\nu(y)}{\int_K \int_K \varphi_0(x)\varphi_0(y) \, d\mu(x) \, d\nu(y)} - \left(\mathbb{E}_{\varphi_0}(\mu) + \mathbb{E}_{\varphi_0}(\nu) \right)^2$$

$$= \frac{\int_K \varphi_2(x) \, d\mu(x)}{\int_K \varphi_0(x) \, d\mu(x)} + 2\mathbb{E}_{\varphi_0}(\mu)\,\mathbb{E}_{\varphi_0}(\nu) + \frac{\int_K \varphi_2(y) \, d\nu(y)}{\int_K \varphi_0(y) \, d\nu(y)} - \left(\mathbb{E}_{\varphi_0}(\mu) + \mathbb{E}_{\varphi_0}(\nu) \right)^2$$

$$= \frac{\int_K \varphi_2(x) \, d\mu(x)}{\int_K \varphi_0(x) \, d\mu(x)} - \mathbb{E}_{\varphi_0}(\mu)^2 + \frac{\int_K \varphi_2(y) \, d\nu(y)}{\int_K \varphi_0(y) \, d\nu(y)} - \mathbb{E}_{\varphi_0}(\nu)^2 = \mathbb{V}_{\varphi_0}(\mu) + \mathbb{V}_{\varphi_0}(\nu).$$

The theorem is proved.   □

Suppose now that K is a commutative hypergroup with the exponential family $\Phi : K \times C^n \to \mathbb{C}$. Let N be a positive integer. For any $k = 1, 2, \ldots, n$, using similar argument and notation as we did in the end of section 1.7, if we differentiate both sides of equation (1.17) N times with respect to the k-th component of the second variable of Φ, then, by Leibniz Rule, we obtain

$$\partial_{k+1}^N \Phi(x * y, \lambda) = \sum_{j=0}^{N} \binom{N}{j} \partial_{k+1}^j \Phi(x, \lambda) \cdot \partial_{k+1}^{N-j} \Phi(y, \lambda)$$

for all x, y in K. This means that the functions $x \mapsto \partial_{k+1}^j \Phi(x, \lambda)$ for $j = 0, 1, \ldots, N$ form a generalized moment function sequence with respect to the exponential $x \mapsto \Phi(x, \lambda)$. In the case of some special hypergroups we will be able to describe the general form of generalized moment function sequences.

In a generalized moment function sequence the function φ_1 plays a special role, as it is shown in the following theorem.

Theorem 1.15. *Let K be a commutative hypergroup, N a positive integer and let $(\varphi_j)_{j=0}^N$ be a generalized moment function sequence. If φ_1 is nonzero, then none of the functions in this sequence is the linear combination of the previous ones. In particular, these functions are linearly independent.*

Proof. Obviously φ_0, as an exponential, is not identically zero. Suppose that $\varphi_1 \neq 0$ is a constant multiple of φ_0, that is, $\varphi_1 = \lambda \varphi_0$ holds with some nonzero complex λ. Then we have for all x, y in K:

$$\varphi_0(x)\varphi_0(y) = \varphi_0(x * y) = \frac{1}{\lambda}\varphi_1(x * y)$$

$$= \frac{1}{\lambda}\varphi_1(x)\varphi_0(y) + \frac{1}{\lambda}\varphi_0(x)\varphi_1(y) = 2\varphi_0(x)\varphi_0(y),$$

which is a contradiction.

Suppose that we have proved our statement for $N = 1, 2, \ldots, k$, where $k \geq 1$ is an integer and now we prove it for $N = k+1$. Assume the contrary, that is, assume that there are complex numbers $\lambda_0, \lambda_1, \ldots, \lambda_k$ such that

$$\varphi_{k+1} = \sum_{j=0}^{k} \lambda_j \varphi_j \tag{1.26}$$

holds. Let x, y be arbitrary in K, then we have

$$\varphi_{k+1}(x * y) = \sum_{j=0}^{k+1} \binom{k+1}{j} \varphi_j(x)\varphi_{k+1-j}(y) \tag{1.27}$$

$$= \sum_{j=0}^{k} \binom{k+1}{j} \varphi_j(x)\varphi_{k+1-j}(y) + \varphi_{k+1}(x)\varphi_0(y).$$

Substitution from equation (1.26) gives

$$\sum_{j=0}^{k} \binom{k+1}{j} \varphi_j(x)\varphi_{k+1-j}(y) + \sum_{j=0}^{k} \lambda_j \varphi_j(x)\varphi_0(y) = \sum_{i=0}^{k} \lambda_i \varphi_i(x*y). \quad (1.28)$$

Here we use again equation (1.19) to get

$$\sum_{j=0}^{k} \left[\binom{k+1}{j} \varphi_{k+1-j}(y) + \lambda_j \varphi_0(y) \right] \varphi_j(x) = \sum_{i=0}^{k} \sum_{j=0}^{i} \lambda_i \varphi_j(x)\varphi_{i-j}(y).$$
$$(1.29)$$

By interchanging the sums on the right hand side we have

$$\sum_{j=0}^{k} \left[\binom{k+1}{j} \varphi_{k+1-j}(y) + \lambda_j \varphi_0(y) \right] \varphi_j(x) = \sum_{j=0}^{k} \sum_{i=j}^{k} \lambda_i \varphi_j(x)\varphi_{i-j}(y),$$
$$(1.30)$$

or

$$\sum_{j=0}^{k} \left[\binom{k+1}{j} \varphi_{k+1-j}(y) + \lambda_j \varphi_0(y) - \sum_{i=j}^{k} \lambda_i \varphi_{i-j}(y) \right] \varphi_j(x) = 0. \quad (1.31)$$

As, by assumption, φ_k is not included in the linear span of the functions $\varphi_0, \varphi_1, \ldots, \varphi_{k-1}$, hence its coefficient in (1.31) must be zero, that is,

$$\binom{k+1}{k} \varphi_{k+1-k}(y) + \lambda_k \varphi_0(y) = \sum_{i=k}^{k} \lambda_i \varphi_{k-k}(y), \qquad (1.32)$$

which implies

$$(k+1)\varphi_1(y) = 0, \qquad (1.33)$$

a contradiction, hence our theorem is proved. $\qquad \square$

1.9 Exponentials and additive functions on a special hypergroup

In this section we present a special hypergroup, which is related to the set of continuous unitary irreducible representations of the group $G = SU(2)$, the *special linear group* in two dimensions. We show how to determine all exponentials and additive functions on this hypergroup and how this problem relates to the problem of solving some special nonlinear recurrence equations. The definition of the underlying hypergroup is taken from [BH95].

If G is a compact topological group, then its dual object $\widehat{G}$ consists of equivalence classes of continuous irreducible representations of G. For any two classes U, V of this type their tensor product can be decomposed into its irreducible components $U_1, U_2, \ldots, U_n$ with the respective multiplicities $m_1, m_2, \ldots, m_n$ (see [HR79]). We define convolution on $\mathcal{M}_b(\widehat{G})$ by

$$\delta_U * \delta_V = \sum_{i=1}^{n} \frac{m_i \, d(U_i)}{d(U) \, d(V)} \, \delta_{U_i} \,,$$

where $d(U)$ denotes the dimension of U. Then $\widehat{G}$ with this convolution and with the discrete topology is a commutative hypergroup.

In the special case of $G = SU(2)$ the dual object $\widehat{G}$ can be identified with the set $\mathbb{N}$ of natural numbers as it is indicated in [BH95]: the set of equivalence classes of continuous unitary irreducible representations of $SU(2)$ is given by $\{T^{(0)}, T^{(1)}, T^{(2)}, \ldots\}$, where $T^{(n)}$ has dimension $n + 1$ and we identify this set with $\mathbb{N}$.

For every m, n in $\mathbb{N}$ the tensor product of $T^{(m)}$ and $T^{(n)}$ is unitary equivalent to

$$T^{(|m-n|)} \bigoplus T^{(|m-n|+2)} \bigoplus \cdots \bigoplus T^{(m+n)} \,.$$

The convolution is given by

$$\delta_m * \delta_n = \sum_{k=|m-n|}^{m+n} {}' \frac{k+1}{(m+1)(n+1)} \delta_k \,, \qquad (1.34)$$

where the prime denotes that only every second term appears in the sum. With this convolution $\mathbb{N}$ becomes a discrete commutative hypergroup and, since all the $T^{(n)}$ are self-conjugate, the hypergroup is in fact Hermitian.

As an illustration of the above ideas we determine all exponential and additive functions on this nontrivial hypergroup.

We recall that the function $M : \mathbb{N} \to \mathbb{C}$ is an exponential if and only if it satisfies

$$M(m)M(n) = \sum_{k=|m-n|}^{m+n} {}' \frac{k+1}{(m+1)(n+1)} M(k) \qquad (1.35)$$

and the function $A : \mathbb{N} \to \mathbb{C}$ is additive if and only if it satisfies

$$A(m) + A(n) = \sum_{k=|m-n|}^{m+n} {}' \frac{k+1}{(m+1)(n+1)} A(k) \qquad (1.36)$$

for all natural numbers m, n. First we describe the solutions of (1.35). For the sake of simplicity we will call the hypergroup described above the *SU(2)-hypergroup*.

Theorem 1.16. *The function $M : \mathbb{N} \to \mathbb{C}$ is an exponential on the $SU(2)$-hypergroup if and only if there exists a complex number λ such that*

$$M(n) = \frac{\sinh[(n+1)\lambda]}{(n+1)\sinh\lambda}$$

holds for each natural number n. (Here $\lambda = 0$ corresponds to the case $M \equiv 1$.)

Proof.   Let $M : \mathbb{N} \to \mathbb{C}$ be a solution of (1.35) and let $f(n) = (n+1)M(n)$ for each n in $\mathbb{N}$. Then we have

$$f(m)f(n) = \sum_{k=|m-n|}^{m+n} {}' f(k)$$

for each m, n in $\mathbb{N}$. With $m = 1$ it follows that f satisfies the following second order homogeneous linear difference equation

$$f(n+2) - f(1)f(n+1) + f(n) = 0$$

for each n in $\mathbb{N}$ with $f(0) = 1$. Let λ be a complex number with $f(1) = 2\cosh\lambda$. Then we have that

$$f(n) = \alpha e^{n\lambda} + \beta e^{-n\lambda}$$

holds for any n in $\mathbb{N}$ with some complex numbers α, β satisfying $\alpha + \beta = 1$. It is easy to see that in this case

$$f(n) = \frac{\sinh[(n+1)\lambda]}{\sinh\lambda}$$

holds for each n in $\mathbb{N}$. Finally, we have

$$M(n) = \frac{\sinh[(n+1)\lambda]}{(n+1)\sinh\lambda}\,.$$

Conversely, it is easy to check that any function M of the given form is an exponential on the $SU(2)$-hypergroup, hence the theorem is proved. $\square$

We recall that, by the terminology introduced in Section 1.6, we can say that the function $\Phi : \mathbb{N} \times \mathbb{C} \to \mathbb{C}$ defined by

$$\Phi(n, \lambda) = \frac{\sinh[(n+1)\lambda]}{(n+1)\sinh\lambda} \tag{1.37}$$

for n in $\mathbb{N}$ and λ in $\mathbb{C}$, $\lambda \neq 0$, with $\Phi(n, 0) = 1$, is an exponential family on the $SU(2)$-hypergroup. One can check easily that

$$\partial_2 \Phi(n, 0) = 0\,,$$

but

$$\partial_2^2 \Phi(n, 0) = \frac{1}{3}n(n+2)\,,$$

hence the function $A_c : \mathbb{N} \to \mathbb{C}$ defined by

$$A_c(n) = \frac{c}{3}n(n+2)$$

for n in $\mathbb{N}$, is an additive function on the $SU(2)$-hypergroup, for each complex number c. The following theorem shows that actually every additive function on the $SU(2)$-hypergroup has this form.

Theorem 1.17. *The function $A : \mathbb{N} \to \mathbb{C}$ is an additive function on the $SU(2)$-hypergroup if and only if there exists a complex number c such that*

$$A(n) = \frac{c}{3}n(n+2)$$

holds for each natural number n.

Proof.   Let $A : \mathbb{N} \to \mathbb{C}$ be a solution of (1.36) and let $f(n) = (n+1)A(n)$ for each n in $\mathbb{N}$. Then we have

$$(n+1)f(m) + (m+1)f(n) = \sum_{k=|m-n|}^{m+n} {}' f(k)$$

for each m, n in $\mathbb{N}$. With $m = 1$ it follows that f satisfies the following second order homogeneous linear difference equation

$$f(n+2) - 2f(n+1) + f(n) = 2c(n+2)$$

for each n in $\mathbb{N}$ with $f(0) = 0$ and $f(1) = 2c$. As the second difference of f is linear it follows that f is a cubic polynomial and simple computation gives that A has the desired form.

Conversely, it is easy to check that any function A of the given form is an additive function on the $SU(2)$-hypergroup, hence the theorem is proved.

$\square$

Chapter 2

Polynomial hypergroups in one variable

2.1 Polynomial hypergroups in one variable

An important special class of Hermitian hypergroups is closely related to orthogonal polynomials.

Let $(a_n)_{n\in\mathbb{N}}$, $(b_n)_{n\in\mathbb{N}}$ and $(c_n)_{n\in\mathbb{N}}$ be real sequences with the following properties: $c_n > 0$, $b_n \geq 0$, $a_{n+1} > 0$ for each n in $\mathbb{N}$, moreover $a_0 = b_0 = 0$ and $a_n + b_n + c_n = 1$ for each n in $\mathbb{N}$. We define the sequence of polynomials $(P_n)_{n\in\mathbb{N}}$ by $P_0(\lambda) = 1$, $P_1(\lambda) = \lambda$ and by the recursive formula

$$\lambda P_n(\lambda) = a_n P_{n-1}(\lambda) + b_n P_n(\lambda) + c_n P_{n+1}(\lambda)$$

for each $n \geq 1$ and λ in $\mathbb{R}$. The following theorem holds (see [BH95]).

Theorem 2.1. *If the sequence of polynomials $(P_n)_{n\in\mathbb{N}}$ satisfies the above conditions, then there exist constants $c(n,l,k)$ for each n,l,k in $\mathbb{N}$ such that*

$$P_n \cdot P_l = \sum_{k=|n-l|}^{n+l} c(n,l,k) P_k$$

holds for each n,l in $\mathbb{N}$.

Proof. By the theorem of Favard (see [Fav39], [Sho36]) the conditions on the sequence of polynomials $(P_n)_{n\in\mathbb{N}}$ imply that there exists a probability measure μ on $[-1,1]$ such that $(P_n)_{n\in\mathbb{N}}$ forms an orthogonal system on $[-1,1]$ with respect to μ. As P_n has degree n, we have

$$P_n \cdot P_l = \sum_{k=0}^{n+l} c(n,l,k) P_k$$

for each n, l in $\mathbb{N}$, where

$$c(n, l, k) = \frac{\int_{-1}^{1} P_k \cdot P_n \cdot P_l \, d\mu}{\int_{-1}^{1} P_k^2 \, d\mu}$$

holds for each n, l, k in $\mathbb{N}$. The orthogonality of $(P_n)_{n \in \mathbb{N}}$ with respect to μ implies $c(n, l, k) = 0$ for $k > n + l$, or $n > l + k$, or $l > n + k$. Hence the statement is proved. $\qquad\qquad\square$

The formula in the theorem is called *linearization formula* and the coefficients $c(n, l, k)$ are called *linearization coefficients*. The recursive formula for the sequence $(P_n)_{n \in \mathbb{N}}$ implies $P_n(1) = 1$ for each n in $\mathbb{N}$, hence we have

$$\sum_{k=|n-l|}^{n+l} c(n, l, k) = 1$$

for each n in $\mathbb{N}$. If the linearization is *nonnegative*, that is, the linearization coefficients are nonnegative: $c(n, l, k) \geq 0$ for each n, l, k in $\mathbb{N}$, then we can define a hypergroup structure on $\mathbb{N}$ by the following rule:

$$\delta_n * \delta_l = \sum_{k=|n-l|}^{n+l} c(n, l, k) \, \delta_k$$

for each n, l in $\mathbb{N}$, with involution as the identity mapping and with e as 0. The resulting discrete Hermitian (hence commutative) hypergroup is called *the polynomial hypergroup associated with the sequence* $(P_n)_{n \in \mathbb{N}}$. We shall denote it by $(\mathbb{N}, (P_n)_{n \in \mathbb{N}})$.

As an example we consider the hypergroup associated with the *Legendre polynomials*. The corresponding recurrence relation is

$$\lambda P_n(\lambda) = \frac{n+1}{2n+1} P_{n+1}(\lambda) + \frac{n}{2n+1} P_{n-1}(\lambda)$$

for each $n \geq 1$ and λ in $\mathbb{R}$. It can easily be seen that the linearization coefficients are nonnegative and the resulting hypergroup associated with the Legendre polynomials is the *Legendre hypergroup*.

Another interesting example for polynomial hypergroups is presented by the *Chebyshev polynomials*. The corresponding recurrence relation in the case of Chebyshev polynomials of the first kind is

$$\lambda T_n(\lambda) = \frac{1}{2} T_{n+1}(\lambda) + \frac{1}{2} T_{n-1}(\lambda)$$

for each $n \geq 1$ and λ in $\mathbb{R}$. Again, it is easy to see that the linearization coefficients are nonnegative and the resulting hypergroup associated with the Chebyshev polynomials of the first kind is the *Chebyshev hypergroup*.

2.2 Exponential and additive functions on polynomial hypergroups

The previous examples about the exponential and additive functions on $D(\theta)$ suggest that there is some hope to describe all exponential and additive functions on different polynomial hypergroups, too. We start with the Chebyshev hypergroup. Suppose that $m : \mathbb{N} \to \mathbb{C}$ is an exponential on the Chebyshev hypergroup, that is, it satisfies

$$m(k * l) = m(k)m(l)$$

for each k, l in $\mathbb{N}$. From the linearization formula it follows easily by induction that

$$T_k(\lambda)T_l(\lambda) = \frac{1}{2}\big(T_{k+l}(\lambda) + T_{|k-l|}(\lambda)\big)$$

holds for each k, l in $\mathbb{N}$ and λ in $\mathbb{C}$. This means that for any function $f : \mathbb{N} \to \mathbb{C}$ we have

$$f(k * l) = \frac{1}{2}\left(f(k + l) + f(|k - l|)\right)$$

for each k, l in $\mathbb{N}$. Consequently, exponentials of the Chebyshev hypergroup are exactly the nonzero solutions of the functional equation

$$m(k + l) + m(|k - l|) = 2m(k)m(l)$$

for each k, l in $\mathbb{N}$. This functional equation is closely related to d'Alembert's functional equation and has been treated – among others – in [Dav01] independently of hypergroups and in [RZ11] on hypergroups. From our consideration it is clear that the functions $k \mapsto T_k(\lambda)$ satisfy this functional equation. In other words, the Chebyshev polynomials evaluated at any complex λ as functions of the subscript present exponential functions on the Chebyshev hypergroup. It turns out that this is true for any polynomial hypergroup. We shall see that the converse is also true: any exponential on a polynomial hypergroup is generated in this way. As different complex values of λ produce different exponentials, this means that the set of all exponentials of a polynomial hypergroup can be identified with the set of all complex numbers.

The following theorem presents a complete description of the exponentials on any polynomial hypergroup (see [BH95], [Szé03]).

Theorem 2.2. *Let K be the polynomial hypergroup associated with the sequence of polynomials $(P_n)_{n \in \mathbb{N}}$. The function $m : \mathbb{N} \to \mathbb{C}$ is an exponential on K if and only if there exists a complex number λ such that*

$$m(k) = P_k(\lambda)$$

holds for each k in $\mathbb{N}$.

Proof. First of all we remark that if a sequence of polynomials $(P_k)_{k \in \mathbb{N}}$ satisfies a recursion of the form

$$P_k(\lambda)P_l(\lambda) = \sum_{j=0}^{k+l} c(k, l, j) P_j(\lambda)$$

with some real or complex coefficients $c(k, l, j)$ for all real λ, then the recursion holds for all complex λ, too. Let λ be a complex number and $m(k) = P_k(\lambda)$ for each k in $\mathbb{N}$. Then by the definition of convolution we have for each k, l in $\mathbb{N}$

$$m(k * l) = \sum_{j=|k-l|}^{k+l} c(k, l, j) m(j)$$

$$= \sum_{j=|k-l|}^{k+l} c(k,l,j)P_j(\lambda) = P_k(\lambda)P_l(\lambda) = m(k)m(l),$$

hence m is an exponential.

Conversely, let m be an exponential on K and we define $\lambda = m(1)$. By the exponential property we have for any positive integer k that

$$\lambda m(k) = m(1)m(k) = m(1 * k) = \sum_{j=k-1}^{k+1} c(k,1,j)m(j)$$

$$= c(k,1,k-1)m(k-1) + c(k,1,k)m(k) + c(k,1,k+1)m(k+1).$$

As the same recursion holds for $k \mapsto P_k(\lambda)$, further $m(0) = 1 = P_0(\lambda)$ and $m(1) = \lambda = P_1(\lambda)$, we infer that $m(k) = P_k(\lambda)$ for each k in $\mathbb{N}$ and the theorem is proved. $\qquad\square$

Applying this result for the Legendre hypergroup we have that the exponential functions in that case are exactly the functions $n \mapsto P_n(\lambda)$ on $\mathbb{N}$, where λ is any complex number and P_n is the n-th Legendre polynomial.

Now we try to describe the additive functions on the Chebyshev hypergroup. By the above considerations we have that the function $a : \mathbb{N} \to \mathbb{C}$ is additive on the Chebyshev hypergroup if and only if it satisfies the functional equation

$$a(k+l) + a(|k-l|) = 2a(k) + 2a(l)$$

for each k, l in $\mathbb{N}$. This functional equation is closely related to the square-norm functional equation. In fact, any solution of this functional equation has the form $a(k) = c \cdot k^2$ with some complex number c. This means that additive functions on the Chebyshev hypergroup are exactly the quadratic functions on $\mathbb{N}$. We can interpret this result in a somewhat surprising manner by observing that $T_n'(1) = n^2$ holds for each n in $\mathbb{N}$, where T_n' is the derivative of the n-th Chebyshev polynomial of the first kind. Consequently, additive functions of the Chebyshev hypergroup have the general form:

$n \mapsto c \cdot T'_n(1)$ with some complex number c. This is a special case of the following remarkable result (see [Szé03]).

Theorem 2.3. *Let K be the polynomial hypergroup associated with the sequence of polynomials $(P_n)_{n \in \mathbb{N}}$. The function $a : \mathbb{N} \to \mathbb{C}$ is an additive function on K if and only if there exists a complex number c such that*

$$a(n) = c\, P'_n(1)$$

holds for each n in $\mathbb{N}$.

Proof.   Suppose first that $a : \mathbb{N} \to \mathbb{C}$ is an additive function on K. Then, by the additive property, we have for any positive integer n that

$$a(n) + a(1) = a(n * 1) = \sum_{k=n-1}^{n+1} c(n, 1, k) a(k)$$

$$= c(n, 1, n-1) a(n-1) + c(n, 1, n) a(n) + c(n, 1, n+1) a(n+1)\,.$$

On the other hand, we have for any positive integer n and for all real λ that

$$\lambda P(\lambda) = c(n, 1, n-1) P_{n-1}(\lambda) + c(n, 1, n) P_n(\lambda) + c(n, 1, n+1) P_{n+1}(\lambda)\,.$$

Differentiating both sides with respect to λ and substituting $\lambda = 1$ we get

$$1 + P'_n(1) = c(n, 1, n-1) P'_{n-1}(1) + c(n, 1, n) P'_n(1) + c(n, 1, n+1) P'_{n+1}(1)$$

for any positive integer n. Multiplying both sides by $a(1)$ we see that the functions $n \mapsto a(n)$ and $n \mapsto a(1) P'_n(1)$ satisfy the same recursion, further $a(0) = a(1) P'_0(1)$ and $a(1) = a(1) P'_1(1)$, hence $a(n) = a(1) P'_n(1)$ for each n in $\mathbb{N}$.

Conversely, we consider the linearization formula

$$P_n(\lambda) P_l(\lambda) = \sum_{k=|n-l|}^{n+l} c(n, l, k) P_k(\lambda)\,,$$

which holds for each n, l in $\mathbb{N}$ and for any real λ. Differentiating both sides with respect to λ and substituting $\lambda = 1$ we get

$$P_n(1)P_l'(1) + P_n'(1)P_l(1) = \sum_{k=|n-l|}^{n+l} c(n, l, k)P_k'(1)$$

for each n, l in $\mathbb{N}$. As $P_n(1) = P_l(1) = 1$ for each n, l in $\mathbb{N}$ this formula shows that the function $n \mapsto P_n'(1)$ is additive, hence $n \mapsto c\, P_n'(1)$ is additive for any complex number c and the theorem is proved. $\qquad\square$

2.3 Moment functions on polynomial hypergroups

In [Gal98] the general form of associated pairs of moment functions satisfying (1.20) has been determined in the case of polynomial hypergroups. Here we restate the corresponding result.

Theorem 2.4. *Let K be the polynomial hypergroup associated with the sequence of polynomials $(P_n)_{n\in\mathbb{N}}$. If (φ_1, φ_2) is an associated pair of real-valued moment functions satisfying (1.20), then there exist real numbers a, b with $b \geq 0$ such that*

$$\varphi_1(n) = aP_n'(1) \qquad and \qquad \varphi_2(n) = a^2 P_n''(1) + bP_n'(1) \qquad (2.1)$$

holds for each n in $\mathbb{N}$. Conversely, if there exist real numbers a, b with $b \geq 0$ satisfying

$$a^2 P_n''(1) + bP_n'(1) \geq a^2 \big(P_n'(1)\big)^2 \qquad (2.2)$$

for each n in $\mathbb{N}$, then the functions (φ_1, φ_2) defined in (2.1) form an associated pair of real-valued moment functions satisfying (1.20).

Proof. If (φ_1, φ_2) is an associated pair of real-valued moment functions satisfying (1.20), then φ_1 is additive, hence it has the form $\varphi_1(n) = aP_n'(1)$ with some real a, by Theorem 2.3. Then, using equation (1.22), we have

$$\varphi_2(n * m) = \varphi_2(n) + 2a^2 P_n'(1)P_m'(1) + \varphi_2(m) \qquad (2.3)$$

for each n, m in $\mathbb{N}$. On the other hand, differentiating the linearization formula for the polynomials P_n twice, substituting $\lambda = 1$ and multiplying by a^2 we have

$$a^2 P_n''(1) + 2a^2 P_n'(1) P_m'(1) + a^2 P_m''(1) = \sum_{k=|n-m|}^{n+m} c(n,m,k) P_k''(1) \qquad (2.4)$$

for each n, m in $\mathbb{N}$. These two equations imply that if

$$\psi(n) = \varphi_2(n) - a^2 P_n''(1)$$

for each n in $\mathbb{N}$, then ψ is additive, hence by Theorem 2.3 again, it follows that $\varphi_2(n) = a^2 P_n''(1) + b P_n'(1)$ holds with some real b for each n in $\mathbb{N}$. Condition (1.20) implies $b \geq 0$ and we have the representation (2.1).

For the converse, using (2.4) we see that the function φ_2 defined in (2.1) satisfies (2.3), which gives (1.22), moreover φ_1 defined in (2.1) is obviously additive, hence (φ_1, φ_2) is an associated pair of real-valued moment functions. Condition (2.2) guarantees (1.20) and the theorem is proved. $\square$

The study of generalized moment functions on hypergroups leads to the study of the system of functional equations (1.19). We remark that a similar system of functional equation on groupoids has been investigated and solved in [Acz77]. As a generalization of the results in [Gal98] here we describe the generalized moment function sequences of order N in the case of polynomial hypergroups (see [OS05]).

Theorem 2.5. *Let K be the polynomial hypergroup associated with the sequence of polynomials $(P_n)_{n \in \mathbb{N}}$. The functions $\varphi_0, \varphi_1, ..., \varphi_N : K \to \mathbb{C}$ form a generalized moment function sequence of order N on K if and only if*

$$\varphi_k(n) = (P_n \circ f)^{(k)}(0)$$

holds for each n in $\mathbb{N}$ and for $k = 0, 1, \ldots, N$, where

$$f(t) = \sum_{j=0}^{N} \frac{c_j}{j!} t^j \qquad (2.5)$$

for each t in $\mathbb{R}$, where c_j is a complex number $(j = 0, 1, \ldots, N)$.

Proof. By the linearization formula we have

$$(P_n \circ f)(t)(P_m \circ f)(t) = \sum_{l=|n-m|}^{n+m} c(n,m,l)(P_l \circ f)(t)$$

for each t in $\mathbb{C}$ and for each n, m in $\mathbb{N}$. Differentiating both sides k times with respect to t and substituting $t = 0$ we have for $k = 0, 1, \ldots, N$

$$\sum_{j=0}^{k} \binom{k}{j} (P_n \circ f)^{(j)}(0)(P_m \circ f)^{(k-j)}(0) = \sum_{l=|n-m|}^{n+m} c(n, m, l)(P_l \circ f)^{(k)}(0)$$

$$= \sum_{l=|n-m|}^{n+m} c(n, m, l)\varphi_k(l) = \varphi_k(n * m),$$

which means that the functions $\varphi_0, \varphi_1, \ldots, \varphi_N : K \to \mathbb{C}$ given above form a generalized moment function sequence of order N on K for any complex numbers c_j $(j = 0, 1, \ldots, N)$.

To prove the converse we suppose that the functions $\varphi_0, \varphi_1, \ldots, \varphi_N :$ $K \to \mathbb{C}$ form a generalized moment function sequence of order N on K and let $c_j = \varphi_j(1)$ for $j = 0, 1, \ldots, N$. Now we define

$$\psi_k(n) = \varphi_k(n) - (P_n \circ f)^{(k)}(0)$$

for $k = 0, 1, \ldots, N$ and for each n in $\mathbb{N}$, where f denotes the function defined by (2.5). We show that the functions $\psi_0, \psi_1, \ldots, \psi_N$ vanish identically on K. For $k = 0$ we have $\psi_0(n) = \varphi_0(n) - P_n(f(0))$ for each n in $\mathbb{N}$. However, as φ_0 is an exponential and $f(0) = \varphi_0(1)$, by Theorem 2.2, it follows $\varphi_0(n) = P_n(\varphi_0(1))$, hence $\psi_0(n) = 0$ for each n in $\mathbb{N}$.

First we observe that $\varphi_k(0) = 0$ for $k = 1, 2, \ldots, N$. Indeed, for $k = 1$ this is obvious from the definition of moment functions and for $k > 1$ it follows by induction. Then we have $\psi_k(0) = 0$ for $k = 1, 2, \ldots, N$. Similarly, $\psi_k(1) = \varphi_k(1) - f^{(k)}(0) = 0$ for $k = 1, 2, \ldots, N$.

Now we proceed by induction on k. Suppose that $\psi_0 = \cdots = \psi_{k-1} = 0$ on K with some $k \geq 1$. We show that ψ_k satisfies a linear homogeneous recurrence relation of the second order and this implies that ψ_k must vanish identically on $K = \mathbb{N}$.

By the linearization formula we have with $m = 1$

$$\sum_{j=0}^{k} (P_n \circ f)^{(j)}(0) f^{(k-j)}(0) = \sum_{l=|n-1|}^{n+1} c(n,1,l)(P_l \circ f)^{(k)}(0)$$

and, by $f^{(k-j)}(0) = \varphi_{k-j}(1)$, this gives

$$\sum_{j=0}^{k} (P_n \circ f)^{(j)}(0) \varphi_{k-j}(1) = \sum_{l=|n-1|}^{n+1} c(n,1,l)(P_l \circ f)^{(k)}(0).$$

We can rewrite this equation in the form

$$\sum_{j=0}^{k-1} (P_n \circ f)^{(j)}(0) \varphi_{k-j}(1) \tag{2.6}$$

$$= \sum_{l=|n-1|}^{n+1} c(n,1,l)(P_l \circ f)^{(k)}(0) - (P_n \circ f)^{(k)}(0) \varphi_0(1).$$

On the other hand, by the definition of the moment functions we have with $m = 1$

$$\sum_{j=0}^{k} \varphi_j(n) \varphi_{k-j}(1) = \varphi_k(n * 1),$$

which can be rewritten in the form

$$\sum_{j=0}^{k-1} \varphi_j(n) \varphi_{k-j}(1) = \varphi_k(n * 1) - \varphi_k(n) \varphi_0(1). \tag{2.7}$$

Subtracting (2.6) from (2.7) we have

$$\psi(n * 1) - \varphi_0(1) \psi(n) = 0,$$

which is the desired recursion and our theorem is proved. $\square$

In the next proposition a given hypergroup structure on the discrete space $\mathbb{N}$ will be identified with the structure of a polynomial hypergroup. Namely, we have the following theorem (see [BH95]).

Theorem 2.6. *Any hypergroup* $(\mathbb{N}, *)$ *such that*

$$\{n+1\} \subseteq supp\,(\delta_1 * \delta_n) \subseteq \{n-1, n, n+1\} \tag{2.8}$$

holds for all n *in* $\mathbb{N}$ *is a polynomial hypergroup.*

Proof. Let e denote the neutral element of $(\mathbb{N}, *)$. From

$$\{e+1\} \subseteq supp\,(\delta_1 * \delta_e) = \{1\}$$

we conclude that $e = 0$. We also note that, by an induction argument, we have

$$supp\,(\delta_m * \delta_n) \subseteq \{n-m, n-m+1, \ldots, n+m\}$$

for all $m \leq n$. In fact for $m = 1$ this is just part of the hypothesis of the proposition. The induction step from m to $m+1$ can be seen from the subsequent chain valid for all m and $n \geq m$

$$supp(\delta_{m+1} * \delta_n)$$

$$\subseteq supp\,(\delta_1 * \delta_m * \delta_n) \subseteq \{1\} * \{n-m, n-m+1, \ldots, n+m\}$$

$$\subseteq \{n-m-1, n-m, \ldots, n+m+1\}.$$

Next we note that $n^- = n$ for all n in $\mathbb{N}$. Otherwise we would have $n^- < n$ for some n in $\mathbb{N}$ which gives

$$0 \in supp\,(\delta_{n^-} * \delta_n) \subseteq \{n - n^-, n - n^- + 1, \ldots, n + n^-\}$$

and this is clearly false. Thus we have that $(\mathbb{N}, *)$ is Hermitian and commutative. Consequently, the set

$$A = \{\chi(1)|\chi \in \widehat{\mathbb{N}}\}$$

is a compact subset of $\mathbb{R}$.

By recursion we now define a sequence $(Q_n)_{n \in \mathbb{N}}$ of polynomials as follows. Put $Q_0 = 1$, $Q_1(x) = x$ and

$$Q_1 Q_n = (\delta_1 * \delta_n)(\{n+1\})Q_{n+1} + (\delta_1 * \delta_n)(\{n\})Q_n + (\delta_1 * \delta_n)(\{n-1\})Q_{n-1}.$$

From the construction of the sequence $(Q_n)_{n \in \mathbb{N}}$ we obtain that

$$\chi(n) = Q_n\big(\chi(1)\big), \tag{2.9}$$

whenever n is in $\mathbb{N}$. Therefore A admits a bijection onto $\widehat{\mathbb{N}}$ and hence it is infinite. In particular, we observe that each χ in $\widehat{\mathbb{N}}$ is determined by its value at 1. For all x in A we have from (2.9) that

$$Q_m(x)Q_n(x) = \sum_{k \in \mathbb{N}} (\delta_m * \delta_n)(\{k\})Q_k(x),$$

whenever m, n are in $\mathbb{N}$. But since A is infinite this relationship extends to all x in $\mathbb{R}$, which shows that $(\mathbb{N}, *)$ is indeed a polynomial hypergroup. $\square$

2.4   Moment functions on the $SU(2)$-hypergroup

In this section we describe the generalized moment functions on the $SU(2)$-hypergroup we studied in Section 1.9. We proceed along the lines of the preceding section.

Theorem 2.7. *Let K denote the $SU(2)$-hypergroup and Φ the exponential family given by (1.37). The functions $\varphi_0, \varphi_1, \ldots, \varphi_N : K \to \mathbb{C}$ form a generalized moment function sequence of order N on K if and only if there exist complex numbers c_j for $j = 1, 2, \ldots, N$ such that*

$$\varphi_k(n) = \frac{d^k}{dt^k}\Phi(n, f(t))(0)$$

holds for each n in $\mathbb{N}$ and for $k = 0, 1, \ldots, N$, where

$$f(t) = \sum_{j=0}^{N} \frac{c_j}{j!} t^j \tag{2.10}$$

for each t in $\mathbb{R}$.

Proof. First we note that, by (1.34), we have for $n \geq 1$

$$\delta_n * \delta_1 = \sum_{k=n-1}^{n+1} {}' \frac{k+1}{2(n+1)} \delta_k = \frac{n}{2(n+1)} \delta_{n-1} + \frac{n+2}{2(n+1)} \delta_{n+1}, \tag{2.11}$$

hence, by Theorem 2.6, K is a polynomial hypergroup, that is, there exists a sequence $(P_n)_{n \in \mathbb{N}}$ of polynomials such that $\deg P_n = n$ for $n = 0, 1, \ldots$, there exists an x_0 in $\mathbb{R}$ such that $P_n(x_0) = 1$ for $n = 0, 1, \ldots$ and

$$P_n(x) P_m(x) = \sum_{k=0}^{\infty} c(m, n, k) P_k(x)$$

holds for each x in $\mathbb{R}$ and m, n in $\mathbb{N}$ with some nonnegative numbers $c(m, n, k)$, further we have

$$\delta_m * \delta_n = \sum_{k=0}^{\infty} c(m, n, k) \delta_k$$

for each m, n in $\mathbb{N}$. Here we shall present this sequence of polynomials.

Our basic observation is that the function $\lambda \mapsto \Phi(n, \lambda)$ is a polynomial of $\cosh \lambda$ of degree n for each n in $\mathbb{N}$. Indeed, by equation (2.11) we have

$$\Phi(n, \lambda) \Phi(1, \lambda) = \frac{n}{2(n+1)} \Phi_{n-1}(\lambda) + \frac{n+2}{2(n+1)} \Phi_{n+1}(\lambda), \tag{2.12}$$

hence, by $\Phi(1, \lambda) = \cosh \lambda$, we have

$$\Phi_{n+1}(\lambda) = \frac{2(n+1)}{n+2} \Phi(n, \lambda) \cosh \lambda - \frac{n}{n+2} \Phi(n-1, \lambda) \tag{2.13}$$

for $n = 1, 2, \ldots$ and for each λ in $\mathbb{C}$. As $\Phi(0, \lambda) = 1$ and $\Phi(1, \lambda) = \cosh \lambda$, our statement follows by induction.

By $\Phi(n, 0) = 1$ for each n in $\mathbb{N}$ we infer that there exists a sequence $(P_n)_{n \in \mathbb{N}}$ of polynomials such that $\deg P_n = n$ for $n = 0, 1, \ldots$, and

$$P_n(\cosh \lambda) = \Phi(n, \lambda) \tag{2.14}$$

holds for each n in $\mathbb{N}$ and λ in $\mathbb{C}$. Clearly $P_0(x) = 1$ and $P_1(x) = x$ for all x in $\mathbb{R}$. Further we have for all m, n in $\mathbb{N}$ and λ in $\mathbb{C}$

$$P_n(\cosh \lambda) P_m(\cosh \lambda) = \Phi(n, \lambda) \Phi(m, \lambda) = \Phi(n * m, \lambda)$$

$$= \sum_{k=|m-n|}^{m+n} {}' \frac{k+1}{(m+1)(n+1)} \Phi(k, \lambda) = \sum_{k=|m-n|}^{m+n} {}' \frac{k+1}{(m+1)(n+1)} P_k(\cosh \lambda),$$

which implies

$$P_n(x) P_m(x) = \sum_{k=|m-n|}^{m+n} {}' \frac{k+1}{(m+1)(n+1)} P_k(x)$$

for each x in $\mathbb{R}$ and m, n in $\mathbb{N}$. This means that K is the polynomial hypergroup associated to the sequence of polynomials $(P_n)_{n \in \mathbb{N}}$. Then, by Theorem 2.5, our statement follows. $\qquad\square$

Chapter 3

Polynomial hypergroups in several variables

3.1 Polynomial hypergroups in several variables

Let K be a *countable* set equipped with the discrete topology and let d be a positive integer. We consider a set $(Q_x)_{x \in K}$ of polynomials in d complex variables. If for any nonnegative integer n the symbol K_n denotes the set of all elements x in K for which the degree of Q_x is not greater than n, then we suppose that the polynomials Q_x with x in K_n form a basis for all polynomials of degree not greater than n. In this case for every x, y in K the product $Q_x Q_y$ admits a unique representation

$$Q_x Q_y = \sum_{w \in K} c(x, y, w) Q_w \qquad (3.1)$$

with some complex numbers $c(x, y, w)$. A hypergroup $(K, *)$ is called a *polynomial hypergroup in d variables* or *d-dimensional polynomial hypergroup* if there exists a family of polynomials $(Q_x)_{x \in K}$ in d complex variables satisfying the above condition and such that the convolution in K is defined by

$$\delta_x * \delta_y(\{w\}) = c(x, y, w)$$

for any x, y, w in K. We say that this polynomial hypergroup is *associated with the family of polynomials* $(Q_x)_{x \in K}$.

It is clear that the polynomial hypergroups in one variable, defined in Section 2.1 represent a special class of this new concept. The above equation (3.1) is a generalization of the linearization formula in Theorem 2.1. It is obvious that any sequence $(p_n)_{n \in \mathbb{N}}$ of polynomials in one variable having the property that for any nonnegative integer n the degree of p_n is exactly n satisfies the above condition in (3.1).

By the conditions on the sequence of polynomials $(Q_x)_{x \in K}$ it follows that there is exactly one element x in K for which Q_x is a nonzero constant. It is easy to see that necessarily $x = e$ is the identity of the hypergroup and $Q_e = 1$. Sometimes it is convenient to identify the element x in K with the polynomial Q_x. Clearly K contains exactly d nonconstant linear polynomials which are linearly independent. If for some $j = 1, 2, \ldots, d$ there exists an x in K for which $Q_x(z_1, z_2, \ldots, z_d) = z_j$, then we say that the polynomial z_j is in K and we denote $z_j^\vee = Q_{x^\vee}$.

3.2 Exponential and additive functions on multivariate polynomial hypergroups

In this section first we characterize the exponential functions on K (see also [BH95]).

Theorem 3.1. *Let K be a d-dimensional polynomial hypergroup generated by the family of polynomials $(Q_x)_{x \in K}$. The function $m : K \to \mathbb{C}$ is an exponential if and only if there exists a λ in $\mathbb{C}^d$ such that*

$$m(x) = Q_x(\lambda) \tag{3.2}$$

holds for all x in K.

Proof.   The linearization formula and the definition of convolution on K shows immediately that any function m of the form given in (3.2) is an exponential on K.

Conversely, suppose that m is any exponential on K. As the polynomials Q_x for x in K_1 form a basis for all linear polynomials, hence there exists a (unique) λ in $\mathbb{C}^d$ such that (3.2) holds for all x in K_1. We show by induction on the degree of Q_x that this is true for any x in K. Suppose that (3.2) holds for any x in K_n and let x be in K_{n+1}. By our assumptions it follows that Q_x has a representation of the form

$$Q_x(\lambda) = \sum_{j=1}^{s} a_j Q_{x_j}(\lambda) Q_{y_j}(\lambda) \tag{3.3}$$

for any λ in $\mathbb{C}^d$ with some complex numbers a_j and with some x_j in K_1 and y_j in K_n $(j = 1, 2, \ldots, s)$, where s is a positive integer. By the definition of the hypergroup structure on K this means that

$$\delta_x = \sum_{j=1}^{s} a_j \delta_{x_j} * \delta_{y_j}$$

holds. Hence we have

$$m(x) = \int_K m \, d\delta_x = \sum_{j=1}^{s} a_j \int_K m \, d(\delta_{x_j} * \delta_{y_j})$$

$$= \sum_{j=1}^{s} a_j m(x_j) m(y_j) = \sum_{j=1}^{s} a_j Q_{x_j}(\lambda) Q_{y_j}(\lambda) = Q_x(\lambda)$$

and our theorem is proved. $\qquad\qquad\qquad\qquad\qquad\qquad\qquad\qquad\qquad\qquad\qquad$ $\square$

This theorem implies that the generalized dual of the d-dimensional polynomial hypergroup can be identified with $\mathbb{C}^d$. Consequently, every polynomial hypergroup admits a normalization in the sense that there exists a λ_0 in $\mathbb{C}^d$ such that $Q_x(\lambda_0) = 1$ holds for any x in K. Indeed, λ_0 is the unique element in $\mathbb{C}^d$ which corresponds to the exponential identically 1. We call λ_0 the *normalizing point* of the hypergroup K. In the case of the polynomial hypergroups of one variable we studied in Section 2.1 the normalizing point was 1.

The next theorem describes additive functions on multivariate polynomial hypergroups.

Theorem 3.2. *Let K be a d-dimensional polynomial hypergroup generated by the family of polynomials $(Q_x)_{x \in K}$ with normalizing point λ_0. The function $a : K \to \mathbb{C}$ is additive if and only if there exist complex numbers c_j for $j = 1, 2, \ldots, d$ such that*

$$a(x) = \sum_{i=1}^{d} c_i \partial_i Q_x(\lambda_0) \tag{3.4}$$

holds for all x in K.

Proof.   By the linearization formula (3.1)

$$Q_x(\lambda)Q_y(\lambda) = \sum_{w \in K} c(x, y, w)Q_w(\lambda)$$

holds for any x, y in K and for any λ in $\mathbb{C}^d$. Applying ∂_i on both sides of this equation and then substituting $\lambda = \lambda_0$ we have for $i = 1, 2, \ldots, d$

$$\partial_i Q_x(\lambda_0) + \partial_i Q_y(\lambda_0) = \sum_{w \in K} c(x, y, w)\partial_i Q_w(\lambda_0),$$

which means that the functions $x \mapsto \partial_i Q_x(\lambda_0)$ are additive for $i = 1, 2, \ldots, d$, hence the function a given in (3.4) is additive for any complex numbers $c_1, c_2, \ldots, c_d$.

For the converse first we observe that the vectors

$$\left(\partial_1 Q_x(\lambda_0), \partial_2 Q_x(\lambda_0), \ldots, \partial_d Q_x(\lambda_0) \right)$$

for x in K_1 and $x \neq e$ are linearly independent, because the polynomials Q_x for x in K_1 form a basis for the linear polynomials in d variables. This implies that the system of linear equations

$$a(x) = \sum_{i=1}^{d} c_i \partial_i Q_x(\lambda_0) \tag{3.5}$$

for x in K_1 with $x \neq e$ has a unique solution $c_1, c_2, \ldots, c_d$. Then (3.5) obviously holds also for $x = e$. We show by induction on n that (3.5) holds for any x in K_n and for any n in $\mathbb{N}$. Supposing that this holds for some n let x be in K_{n+1}. Similarly, like in the proof of the previous theorem, we have that Q_x has a representation in the form (3.3) for all λ in $\mathbb{C}^d$ with some complex numbers a_j and with some x_j in K_1 and y_j in K_n ($j = 1, 2, \ldots, s$), where s is a positive integer, which means that

$$\delta_x = \sum_{j=1}^{s} a_j \delta_{x_j} * \delta_{y_j}$$

holds. On the other hand, applying ∂_i on (3.3) and substituting $\lambda = \lambda_0$ we have for $i = 1, 2, \ldots, d$

$$\partial_i Q_x(\lambda_0) = \sum_{j=1}^{s} a_j \big(\partial_i Q_{x_j}(\lambda_0) + \partial_i Q_{y_j}(\lambda_0) \big).$$

Finally we obtain

$$a(x) = \int_K a \, d\delta_x = \sum_{j=1}^{s} a_j \int_K a(t) \, d(\delta_{x_j} * \delta_{y_j})(t)$$

$$= \sum_{j=1}^{s} a_j \big(a(x_j) + a(y_j) \big) = \sum_{j=1}^{s} a_j \sum_{i=1}^{d} c_i \big(\partial_i Q_{x_j}(\lambda_0) + \partial_i Q_{y_j}(\lambda_0) \big)$$

$$= \sum_{i=1}^{d} c_i \sum_{j=1}^{s} a_j \big(\partial_i Q_{x_j}(\lambda_0) + \partial_i Q_{y_j}(\lambda_0) \big) = \sum_{i=1}^{d} c_i \partial_i Q_x(\lambda_0)$$

and our theorem is proved. □

3.3 Moment function sequences on multivariate polynomial hypergroups

Here we generalize the results in Section 2.3 by characterizing moment functions on multivariate polynomial hypergroups (see [OS04]).

Theorem 3.3. *Let K be a d-dimensional polynomial hypergroup generated by the family of polynomials $(Q_x)_{x \in K}$. The functions φ_0, $\varphi_1, \ldots, \varphi_N : K \to \mathbb{C}$ form a generalized moment function sequence of order N on K if and only if*

$$\varphi_k(x) = (Q_x \circ f)^{(k)}(0) \tag{3.6}$$

holds for all n in $\mathbb{N}$ and for $k = 0, 1, \ldots, N$, where the function $f = (f_1, f_2, \ldots, f_d) : \mathbb{R} \to \mathbb{C}$ is such that

$$f_i(t) = \sum_{j=0}^{N} \frac{c_{i,j}}{j!} t^j$$

for all t in $\mathbb{R}$, where $c_{i,j}$ is a complex number for each $i = 1, 2, \ldots, d$; $j = 1, 2, \ldots, N$.

Proof.   Let φ_k denote the function defined by (3.6) for $k = 0, 1, \ldots, N$. By the linearization formula we have

$$(Q_x \circ f)(t)(Q_y \circ f)(t) = \sum_{w \in K} c(x, y, w)(Q_w \circ f)(t)$$

for any t in $\mathbb{C}$ and for all x, y in K. Differentiating both sides k times with respect to t and substituting $t = 0$ we have for $k = 0, 1, \ldots, N$ and for any x, y in K

$$\sum_{j=0}^{k} \binom{k}{j} \varphi_j(x) \varphi_{k-j}(y)$$

$$= \sum_{j=0}^{k} \binom{k}{j} (Q_x \circ f)^{(j)}(0)(Q_y \circ f)^{(k-j)}(0) = \sum_{w \in K} c(x, y, w)(Q_w \circ f)^{(k)}(0)$$

$$= \sum_{w \in K} c(x, y, w) \varphi_k(w) = \varphi_k(x * y),$$

which means that the functions $\varphi_0, \varphi_1, \ldots, \varphi_N : K \to \mathbb{C}$ given above form a generalized moment sequence of order N on K for any complex numbers $c_{i,j}$ $(i = 1, 2, \ldots, d; \ j = 1, 2, \ldots, N)$.

To prove the converse statement we suppose now that the functions $\varphi_0, \varphi_1, \ldots, \varphi_N : K \to \mathbb{C}$ form a generalized moment function sequence of order N on K. As φ_0 is an exponential, we have that $\varphi_0(x) = Q_x(\lambda)$ holds for all x in K with some λ in $\mathbb{C}^d$. We have seen in the proof of the previous theorem, that the vectors

$$\left(\partial_1 Q_x(\lambda), \partial_2 Q_x(\lambda), \ldots, \partial_d Q_x(\lambda) \right)$$

for x in K_1 and $x \neq e$ are linearly independent, consequently, for any fixed $j = 1, 2, \ldots, N$ the system of linear equations

$$\varphi_j(x) = \sum_{i=1}^{d} c_{i,j} \partial_i Q_x(\lambda)$$

for x in K_1 with $x \neq e$ has a unique solution $c_{i,j}$ $(i = 1, 2, \ldots, d)$. Then we define $f = (f_1, f_2, \ldots, f_d)$ by $\lambda = (c_{1,0}, c_{2,0}, \ldots, c_{d,0})$ and

$$f_i(t) = \sum_{j=0}^{N} \frac{c_{i,j}}{j!} t^j$$

for all t in $\mathbb{C}$ and for $i = 1, 2, \ldots, d$, further let

$$\psi_k(x) = \varphi_k(x) - (Q_x \circ f)^{(k)}(0)$$

for $k = 0, 1, \ldots, N$ and for all x in K. We show that the functions $\psi_0, \psi_1, \ldots, \psi_N$ vanish identically on K. For $k = 0$ we have $\psi_0(x) = \varphi_0(x) - Q_x(f(0))$ for all n in $\mathbb{N}$. However, as $f(0) = \lambda$, it follows immediately from the choice of λ that $\varphi_0(x) = Q_x(f(0))$, hence $\psi_0(x) = 0$ for all x in K.

From the equation of the moment functions it follows by induction on k that $\varphi_k(e) = 0$ for $k = 1, 2, \ldots, N$, consequently, we have $\psi_k(e) = 0$ for $k = 0, 1, \ldots, N$. On the other hand, for any x in K_1 the polynomial Q_x is linear, hence

$$(Q_x \circ f)^{(k)}(0) = \sum_{i=1}^{d} \partial_i Q_x(f(0)) f_i^{(k)}(0) = \sum_{i=1}^{d} \partial_i Q_x(\lambda) c_{i,k} = \varphi_k(x)$$

holds for $k = 1, 2, \ldots, N$, whenever $x \neq e$. This means that $\psi_k(x) = 0$ for any x in K_1 and for $k = 0, 1, \ldots, N$.

Now we proceed by induction. Suppose that we have proved $\psi_k(x) = 0$ for $k = 0, 1, \ldots, N$ and for any x in K_n and let x be arbitrary in K_{n+1}. By the same argument that we used in the proof of the previous theorems we have that Q_x has a representation in the form (3.3) for all λ in $\mathbb{C}^d$ with some complex numbers a_j and with some x_j in K_1 and y_j in K_n $(j = 1, 2, \ldots, s)$, where s is a positive integer, which means that

$$\delta_x = \sum_{j=1}^{s} a_j \delta_{x_j} * \delta_{y_j}$$

holds. Consequently, we have

$$\varphi_k(x) = \sum_{j=1}^{s} a_j \varphi_k(x_j * y_j)$$

for $k = 1, 2, \ldots, N$. On the other hand, differentiating (3.3) k times and substituting $t = 0$ we have

$$\left(Q_x \circ f\right)^{(k)}(0) = \sum_{j=1}^{s} a_j \sum_{l=0}^{k} \binom{k}{l} \left(Q_{x_j} \circ f\right)^{(l)}(0) \left(Q_{y_j} \circ f\right)^{(k-l)}(0)$$

$$= \sum_{j=1}^{s} a_j \sum_{l=0}^{k} \binom{k}{l} \varphi_l(x_j) \varphi_{k-l}(y_j) = \sum_{j=1}^{s} a_j \varphi_k(x_j * y_j) = \varphi_k(x),$$

which means that $\psi_k(x) = 0$ for $k = 0, 1, \ldots, N$. This completes the proof.

$\square$

Chapter 4

Sturm–Liouville hypergroups

4.1 Sturm–Liouville functions

Sturm–Liouville hypergroups represent another important class of hyper-groups, which arise from Sturm–Liouville boundary value problems on the nonnegative reals. In order to build up the Sturm–Liouville opera-tor, which is basic to the construction of hypergroups we introduce the Sturm–Liouville functions. For further details see [BH95]. A special im-portant class of Sturm–Liouville hypergroups is formed by the *Chébli-Trimèche hypergroups* (see [Tri05a], [Tri05b], [BX00a], [NRT98], [JT98], [BX97], [LT95]). In what follows $\mathbb{R}_0$ denotes the set of nonnegative real numbers.

The continuous function $A : \mathbb{R}_0 \to \mathbb{R}$ is called a *Sturm–Liouville func-tion*, if it is positive and continuously differentiable on the positive reals. Different assumptions on A can be found in [BH95], which lead to the de-sired Sturm–Liouville problem. For a given Sturm–Liouville function A we define the *Sturm–Liouville operator L_A* by

$$L_A f = -f'' - \frac{A'}{A} f' \,,$$

where f is a twice continuously differentiable real function on the positive reals. Using L_A we define the differential operator l by

$$l[u](x, y) = (L_A)_x u(x, y) - (L_A)_y u(x, y)$$

$$= -\partial_1^2 u(x, y) - \frac{A'(x)}{A(x)} \partial_1 u(x, y) + \partial_2^2 u(x, y) + \frac{A'(y)}{A(y)} \partial_2 u(x, y) \,,$$

where u is twice continuously differentiable for all positive reals x, y. Here $(L_A)_x$ and $(L_A)_y$ indicates that L_A operates on functions depending on x or y, respectively.

A hypergroup on $\mathbb{R}_0$ is called a *Sturm–Liouville hypergroup* if there exists a Sturm–Liouville function A such that given any real-valued function f on $\mathbb{R}_0$ that is the restriction of an even nonnegative $\mathcal{C}^\infty$-function on $\mathbb{R}$ the function u_f defined by

$$u_f(x, y) = \int_{\mathbb{R}_0} f \, d(\delta_x * \delta_y)$$

for all positive x, y is twice continuously differentiable and satisfies the partial differential equation

$$l[u_f] = 0$$

with $\partial_2 u_f(x, 0) = 0$ for all positive x. In other words, u_f is a solution of the Cauchy-problem

$$\partial_1^2 u(x, y) + \frac{A'(x)}{A(x)} \partial_1 u(x, y) = \partial_2^2 u(x, y) + \frac{A'(y)}{A(y)} \partial_2 u(x, y),$$

$$\partial_2 u_f(x, 0) = 0$$

for all positive x, y. From general properties of one-dimensional hypergroups given in [BH95] it follows that $u_f(y, 0) = u_f(0, y) = f(y)$ and $\partial_1 u_f(0, y) = 0$, whenever y is a positive real number. In other words, u_f is the unique solution of the boundary value problem

$$\partial_1^2 u(x, y) + \frac{A'(x)}{A(x)} \, \partial_1 u(x, y) \;=\; \partial_2^2 u(x, y) + \frac{A'(y)}{A(y)} \, \partial_2 u(x, y)$$

$$\partial_1 u(0, y) = 0, \qquad \partial_2 u(x, 0) = 0, \tag{4.1}$$

$$u(x, 0) = f(x), \qquad u(0, y) = f(y)$$

for all positive x, y. As this boundary value problem uniquely defines u_f for any f in question, we may consider it the *boundary value problem defining the Sturm–Liouville hypergroup.*

The following theorem is important (see [BH95], Proposition 3.4.2).

Theorem 4.1. *If $\mathbb{R}_0$ with its euclidean topology is the underlying space of a hypergroup, then its neutral element is 0 and the hypergroup is Hermitian (hence commutative).*

If a Sturm–Liouville hypergroup structure is given on $\mathbb{R}_0$ by the Sturm–Liouville function A, then we denote it by $(\mathbb{R}_0, A)$. If the Sturm–Liouville function A satisfies

$$\frac{A'(x)}{A(x)} = \frac{\alpha_0}{x} + \alpha_1(x) \tag{4.2}$$

for all $x \neq 0$ in a neighborhood of 0 with $\alpha_0 > 0$ such that α_1 is an odd C^∞-function on $\mathbb{R}$ and the function $\frac{A'}{A}$ is nonnegative and decreasing, further A is increasing with $\lim_{x \to +\infty} A(x) = +\infty$, then A is called a *Chébli–Trimèche function* and the corresponding Sturm–Liouville hypergroup is called a *Chébli–Trimèche hypergroup*. Special cases are represented by the *Bessel–Kingman hypergroups* with $A(0) = 0$ and

$$A(x) = x^\alpha$$

for all positive x and some $\alpha > 0$, further the *hyperbolic hypergroups* where $A(0) = 0$ and

$$A(x) = \sinh^a x$$

holds for all positive x and for some $a > 0$.

If the Sturm–Liouville function A is twice continuously differentiable on the positive reals and satisfies (4.2), where $\alpha_0 = 0$ and α_1 is continuously differentiable on the positive reals, then A is called a *Levitan function* and the corresponding Sturm–Liouville hypergroup is called a *Levitan hypergroup*. Special cases are represented by the *cosh hypergroup* where

$$A(x) = \cosh^2 x$$

for all nonnegative x and the *square hypergroup* where

$$A(x) = (1 + x)^2$$

for all nonnegative x (see [Zeu92]). For more about these hypergroups and their applications see [BH95].

4.2 Exponentials and additive functions on Sturm–Liouville hypergroups

In this section we study exponentials on Sturm–Liouville hypergroups. It turns out that the exponentials on these hypergroups can be described using the eigenfunctions of the corresponding Sturm–Liouville operators, that is, by the solutions of boundary value problems.

Theorem 4.2. *Let* $K = (\mathbb{R}_0, A)$ *be a Sturm–Liouville hypergroup corresponding to the Sturm–Liouville function* $A : \mathbb{R}_0 \to \mathbb{R}$. *Then the continuous function* $m : \mathbb{R}_0 \to \mathbb{C}$ *is an exponential on* K *if and only if it is* C^∞ *on the positive reals and there exists a complex number* λ *such that*

$$m''(x) + \frac{A'(x)}{A(x)} m'(x) = \lambda m(x), \qquad m(0) = 1, \qquad m'(0) = 0 \qquad (4.3)$$

holds for any positive x. *Moreover,* m *is real-valued if and only if* λ *is real.*

Proof. First suppose that the function $m : \mathbb{R}_0 \to \mathbb{C}$ is C^∞ on the positive reals and it satisfies the given boundary value problem. Then we have that the function

$$m(x * y) = \int_0^\infty m(t) \, d(\delta_x * \delta_y)(t)$$

and also the function $(x, y) \to m(x)m(y)$ is a solution of the boundary value problem defining the hypergroup, hence they are equal and m is an exponential.

Conversely, suppose that $m : \mathbb{R}_0 \to \mathbb{C}$ is an exponential on the hypergroup K. Then the function $u_m(x, y) = m(x)m(y)$ is a solution of the boundary value problem defining the hypergroup, hence we obtain

$$\left(m''(x) + \frac{A'(x)}{A(x)} m'(x) \right) m(y) = \left(m''(y) + \frac{A'(y)}{A(y)} m'(y) \right) m(x)$$

holds for each positive x, y and there exists a complex λ with

$$m''(x) + \frac{A'(x)}{A(x)} m'(x) = \lambda m(x)$$

for all positive x. The relations $m(0) = 1$ and $m'(0) = 0$ are immediate consequences of the fact that m is an exponential and the neutral element of the hypergroup is zero. As any Sturm–Liouville hypergroup is Hermitian, the last statement of the theorem follows. $\qquad\square$

Hence if the Sturm–Liouville hypergroup $K = (\mathbb{R}_0, A)$ is given, then for each complex number λ there exists a unique exponential m_λ of K satisfying the second order initial value problem (4.3). Conversely, by the theorem, every exponential on K has this form. Hence we can define an *exponential family* $M : \mathbb{R}_0 \times \mathbb{C} \to \mathbb{C}$ with the property that the function $x \mapsto M(x, \lambda)$ is an exponential of K for each complex λ and for each exponential m of K there exists a unique complex λ such that $m(x) = M(x, \lambda)$ holds for every x in $\mathbb{R}_0$.

We obtain the following theorem in a similar way.

Theorem 4.3. *Let $K = (\mathbb{R}_0, A)$ be a Sturm–Liouville hypergroup corresponding to the Sturm–Liouville function $A : \mathbb{R}_0 \to \mathbb{R}$. Then the continuous function $a : \mathbb{R}_0 \to \mathbb{C}$ is an additive function on K if and only if it is C^∞ on the positive reals and there exists a complex number λ such that*

$$a''(x) + \frac{A'(x)}{A(x)} a'(x) = \lambda, \qquad a(0) = 0, \qquad a'(0) = 0 \qquad (4.4)$$

holds for any positive x. Moreover, a is real-valued if and only if λ is real.

Proof. Supposing that the function $a : \mathbb{R}_0 \to \mathbb{C}$ is C^∞ on the positive reals and it satisfies the given boundary value problem we have that the function

$$a(x * y) = \int_0^\infty a(t) \, d(\delta_x * \delta_y)(t)$$

and also the function $(x, y) \to a(x) + a(y)$ is a solution of the boundary value problem defining the hypergroup, hence they are equal and a is additive.

Conversely, suppose that $a : \mathbb{R}_0 \to \mathbb{C}$ is an additive function on the hypergroup K. Then the function $u_a(x, y) = a(x) + a(y)$ is C^∞ and it is a solution of the boundary value problem defining the hypergroup, hence we obtain

$$a''(x) + \frac{A'(x)}{A(x)} a'(x) = a''(y) + \frac{A'(y)}{A(y)} a'(y)$$

holds for each positive x, y, hence there exists a complex λ with

$$a''(x) + \frac{A'(x)}{A(x)} a'(x) = \lambda$$

for all positive x. The relations $a(0) = 0$ and $a'(0) = 0$ are immediate consequences of the fact that a is additive and the neutral element of the hypergroup is zero. As any Sturm–Liouville hypergroup is Hermitian, the last statement of the theorem follows, like in the previous theorem. $\quad\square$

It is obvious that the unique solution a_λ of the boundary value problem (4.4) is λa_1, where a_1 is the unique solution of (4.4) with $\lambda = 1$. This means that all additive functions of a Sturm–Liouville hypergroup are constant multiples of a fixed nonzero additive function. We call a_1 the *generating additive function* of the Sturm–Liouville hypergroup $(\mathbb{R}_0, A)$.

It turns out that the boundary value problem (4.4) can be solved explicitly. Namely, we have the following theorem.

Theorem 4.4. *Let* $K = (\mathbb{R}_0, A)$ *be a Sturm–Liouville hypergroup corresponding to the Sturm–Liouville function* $A : \mathbb{R}_0 \to \mathbb{R}$. *Then the generating additive function of the hypergroup* K *is given by*

$$a_1(x) = \int_0^x \int_0^y \frac{A(t)}{A(y)}\, dt\, dy \tag{4.5}$$

for each nonnegative x. *Hence any additive function of the hypergroup* K *is given by*

$$a_\lambda(x) = \lambda \int_0^x \int_0^y \frac{A(t)}{A(y)}\, dt\, dy \tag{4.6}$$

for each nonnegative x.

Proof. The proof is obvious using standard methods from the theory of linear differential equations. Another way of proving the statement is direct verification and using the uniqueness theorem. $\quad\square$

As an illustration we compute the additive functions on the Bessel–Kingman hypergroup, which is a special Chébli–Trimèche hypergroup. Here $A(x) = x^\alpha$ for all nonnegative x with some positive number α. In this case we have

$$a_1(x) = \int_0^x \int_0^y \frac{t^\alpha}{y^\alpha}\, dt\, dy = \frac{x^2}{2(\alpha + 1)}$$

and

$$a_\lambda(x) = \lambda \int_0^x \int_0^y \frac{t^\alpha}{y^\alpha}\, dt\, dy = \frac{\lambda x^2}{2(\alpha+1)}$$

for each nonnegative x and complex number λ.

Another example is given here for a special Levitan hypergroup, the square hypergroup, where $A(x) = (1+x)^2$ for all nonnegative x. From the above formulas we have

$$a_1(x) = \int_0^x \int_0^y \frac{(1+t)^2}{(1+y)^2}\, dt\, dy = \frac{x^3 + 3x^2}{6(x+1)},$$

$$a_\lambda(x) = \lambda \int_0^x \int_0^y \frac{(1+t)^2}{(1+y)^2}\, dt\, dy = \frac{\lambda(x^3 + 3x^2)}{6(x+1)},$$

for each nonnegative x and complex number λ.

It is not an easy job to find explicit representations for exponential functions on Sturm–Liouville hypergroups. For instance, in the case of the Bessel–Kingman hypergroup, where $A(x) = x^\alpha$ for all nonnegative x with some positive number α we have to solve the Cauchy-problem

$$m''(x) + \frac{\alpha}{x} m'(x) = \lambda m(x), \qquad m(0) = 1, \qquad m'(0) = 0$$

holds for any positive x, where λ is a complex parameter. We can do this in the trivial case $\lambda = 0$ and the corresponding exponential is the identically 1 function.

4.3 Moment functions on Sturm–Liouville hypergroups

Let $K = (\mathbb{R}_0, A)$ be a Sturm–Liouville hypergroup. In this section we describe all generalized moment functions defined on K.

Theorem 4.5. *Let $K = (\mathbb{R}_0, A)$ be the Sturm–Liouville hypergroup corresponding to the Sturm–Liouville function A and let N be a positive integer. The continuous functions $f_k : \mathbb{R}_0 \to \mathbb{C}$ ($k = 0, 1, \ldots, N$) form a sequence of generalized moment functions on the hypergroup K if and only if they are C^∞ and there are complex numbers c_k for $k = 0, 1, \ldots, N$ such that*

$$f_0''(x) + \frac{A'(x)}{A(x)} f_0'(x) = c_0\, f_0(x), \qquad f_0(0) = 1, \qquad f_0'(0) = 0 \qquad (4.7)$$

and

$$f_k''(x) + \frac{A'(x)}{A(x)} f_k'(x) = \sum_{j=0}^{k} \binom{k}{j} c_j \, f_{k-j}(x), \qquad f_k(0) = 0, \qquad f_k'(0) = 0$$

(4.8)

holds for each positive x and for $k = 1, 2, \ldots, N$.

Proof. First we proof the sufficiency of the condition. If the given functions $f_k : \mathbb{R}_0 \to \mathbb{C}$ ($k = 0, 1, \ldots, N$) satisfy the conditions (4.7) and (4.8), then f_0 is an exponential function and hence $f_0(x * y) = f_0(x)f_0(y)$ holds for all nonnegative numbers x and y. We show that equation (1.19) holds for all $k = 1, \ldots, N$, namely, the function

$$h(x, y) = \sum_{j=0}^{k} \binom{k}{j} f_j(x) f_{k-j}(y)$$

is a solution of the differential equation in (4.1). Indeed, the differential equation (4.1) is equivalent to

$$\sum_{j=0}^{k} \binom{k}{j} f_j''(x) f_{k-j}(y) + \frac{A'(x)}{A(x)} \sum_{j=0}^{k} \binom{k}{j} f_j'(x) f_{k-j}(y)$$

$$= \sum_{j=0}^{k} \binom{k}{j} f_j(x) f_{k-j}''(y) + \frac{A'(y)}{A(y)} \sum_{j=0}^{k} \binom{k}{j} f_j(x) f_{k-j}'(y),$$

which is equivalent to

$$\sum_{j=0}^{k} \binom{k}{j} \left(f_j''(x) + \frac{A'(x)}{A(x)} f_j'(x) \right) f_{k-j}(y)$$

$$= \sum_{j=0}^{k} \binom{k}{j} \left(f_{k-j}''(y) + \frac{A'(y)}{A(y)} f_{k-j}'(y) \right) f_j(x),$$

that is, to

$$\sum_{j=0}^{k} \binom{k}{j} \left(\sum_{t=0}^{j} \binom{j}{t} c_t f_{j-t}(x) \right) f_{k-j}(y)$$

$$= \sum_{j=0}^{k} \binom{k}{j} \left(\sum_{s=0}^{k-j} \binom{k-j}{s} c_s f_{k-j-s}(y) \right) f_j(x) \, .$$

But this equation holds true, since by choosing $l = j + s$, the right hand side is equal to

$$\sum_{l=0}^{k} \sum_{s=0}^{l} \binom{k}{l-s} \binom{k-(l-s)}{s} c_s f_{k-l}(y) f_{l-s}(x) \, ,$$

which is obviously equal to the left hand side. Moreover, the boundary value conditions in (4.1) are also satisfied, as

$$\partial_1 h(0,y) = \sum_{j=0}^{k} f_j'(0) f_{k-j}(y) = 0$$

and

$$h(0,y) = \sum_{j=0}^{k} f_j(0) f_{k-j}(y) = f_k(y)$$

and similarly $\partial_2 h(x,0) = 0$ and $h(x,0) = f_k(x)$, hence h is actually the unique solution of the boundary value problem (4.1), which implies that $h(x,y) = f(x * y)$.

Conversely, suppose that the given continuous functions $f_k : \mathbb{R}_0 \to \mathbb{C}$ ($k = 0, 1, \ldots, N$) form a generalized moment sequence of order N. Then, by definition, f_0 is an exponential, a C^∞ function and the conditions in (4.7) are satisfied. Now we proceed by induction and assume that (4.8) holds for the C^∞-functions $f_0, f_1, \ldots, f_k$, with some positive integer $k < N$. We have

$$f_{k+1}(x * y) = \sum_{j=0}^{k+1} \binom{k+1}{j} f_j(x) f_{k+1-j}(y) \tag{4.9}$$

and, by the definition of the hypergroup, this implies that

$$\sum_{j=0}^{k+1} \binom{k+1}{j} f_j''(x) f_{k+1-j}(y) + \frac{A'(x)}{A(x)} \sum_{j=0}^{k+1} \binom{k+1}{j} f_j'(x) f_{k+1-j}(y)$$

$$= \sum_{j=0}^{k+1} \binom{k+1}{j} f_j(x) f_{k+1-j}''(y) + \frac{A'(y)}{A(y)} \sum_{j=0}^{k+1} \binom{k+1}{j} f_j(x) f_{k+1-j}'(y).$$

Rearranging the terms and using the induction hypothesis we have

$$\left(f_{k+1}''(x) + \frac{A'(x)}{A(x)} f_{k+1}'(x) \right) f_0(y)$$

$$+ \sum_{j=0}^{k} \binom{k+1}{j} \left(\sum_{t=0}^{j} \binom{j}{t} c_t f_{j-t}(x) \right) f_{k+1-j}(y)$$

$$= \left(f_{k+1}''(y) + \frac{A'(y)}{A(y)} f_{k+1}'(y) \right) f_0(x)$$

$$+ \sum_{j=1}^{k+1} \binom{k+1}{j} \left(\sum_{t=0}^{k+1-j} \binom{k+1-j}{t} c_t f_{k+1-j-t}(y) \right) f_j(x),$$

therefore

$$\left(f_{k+1}''(x) + \frac{A'(x)}{A(x)} f_{k+1}'(x) \right) f_0(y) + \sum_{j=0}^{k} \binom{k+1}{j} c_j f_0(x) f_{k+1-j}(y)$$

$$+ \sum_{j=1}^{k} \binom{k+1}{j} \left(\sum_{t=0}^{j-1} \binom{j}{t} c_t f_{j-t}(x) \right) f_{k+1-j}(y)$$

$$= \left(f_{k+1}''(y) + \frac{A'(y)}{A(y)} f_{k+1}'(y) \right) f_0(x) + \sum_{j=1}^{k+1} \binom{k+1}{j} c_{k+1-j} f_0(y) f_j(x)$$

$$+ \sum_{j=1}^{k} \binom{k+1}{j} \left(\sum_{t=0}^{k-j} \binom{k+1-j}{t} c_t f_{k+1-j-t}(y) \right) f_j(x).$$

It is easy to see that the last terms on the two sides are equal:

$$\sum_{j=1}^{k} \binom{k+1}{j} \left(\sum_{t=0}^{j-1} \binom{j}{t} c_t f_{j-t}(x) \right) f_{k+1-j}(y)$$

$$= \sum_{t=0}^{k-1} \sum_{s=1}^{k-t} \binom{k+1}{s} \binom{k+1-s}{t} c_t f_{k+1-s-t}(x) f_s(y)$$

$$= \sum_{t=0}^{k-1} \sum_{s=1}^{k-t} \binom{k+1}{s+t} \binom{s+t}{t} c_t f_s(y) f_{k+1-s-t}(x)$$

$$= \sum_{j=1}^{k} \binom{k+1}{j} \left(\sum_{t=0}^{k-j} \binom{k+1-j}{t} c_t f_{k+1-j-t}(y) \right) f_j(x).$$

This means that

$$\left(f''_{k+1}(x) + \frac{A'(x)}{A(x)} f'_{k+1}(x) - \sum_{j=1}^{k+1} \binom{k+1}{j} c_{k+1-j} f_j(x) \right) f_0(y)$$

$$= \left(f''_{k+1}(y) + \frac{A'(y)}{A(y)} f'_{k+1}(y) - \sum_{j=0}^{k} \binom{k+1}{j} c_j f_{k+1-j}(y) \right) f_0(x)$$

holds for each positive x and y, hence there exists a complex number c_{k+1} such that

$$f''_{k+1}(x) + \frac{A'(x)}{A(x)} f'_{k+1}(x) - \sum_{j=1}^{k+1} \binom{k+1}{j} c_{k+1-j} f_j(x) = c_{k+1} f_0(x).$$

As a consequence of (4.9) we also have $f_{k+1}(0) = 0$ and due to

$$0 = \sum_{j=0}^{k+1} \binom{k+1}{j} f_j(x) f'_{k+1-j}(0) = f_0(x) f'_{k+1}(0)$$

we get that $f'_{k+1}(0) = 0$. Hence (4.8) holds for $k+1$, f_{k+1} is obviously C^∞ and the theorem is proved by induction. $\square$

Theorem 4.6. *Let $K = (\mathbb{R}_0, A)$ be the Sturm–Liouville hypergroup corresponding to the Sturm–Liouville function A with the exponential family φ and let N be a positive integer. The continuous functions $f_k : \mathbb{R}_0 \to \mathbb{C}$ ($k = 0, 1, \ldots, N$) form a sequence of generalized moment functions of order N on the hypergroup K if and only if there are complex numbers $c_0, c_1, \ldots, c_N$ such that*

$$f_k(x) = \partial_t^k \varphi\big(x, f(t)\big)\big|_{t=0} \qquad (4.10)$$

holds for each x in $\mathbb{R}_0$, where

$$f(t) = \sum_{j=0}^{N} c_j \frac{t^j}{j!}$$

for each t in $\mathbb{R}$.

Proof. Let k be in $\{0, 1 \ldots, N\}$ and let f be the function given in the theorem. If we take $\lambda = f(t)$ in (4.3) and differentiate the equation k-times with respect to t we get that

$$\partial_x^2 \partial_t^k \varphi\big(x, f(t)\big) + \frac{A'(x)}{A(x)} \partial_x \partial_t^k \varphi\big(x, f(t)\big) = \sum_{j=0}^{k} \binom{k}{j} f^{(j)}(t) \partial_t^{k-j} \varphi\big(x, f(t)\big).$$

$$(4.11)$$

Taking $t = 0$ we have for $f_k(x) = \partial_t^k \varphi\big(x, f(t)\big)\big|_{t=0}$ the following equation

$$f_k''(x) + \frac{A'(x)}{A(x)} f_k'(x) = \sum_{j=0}^{k} \binom{k}{j} c_j \, f_{k-j}(x),$$

furthermore $f_0(0) = 1$, $f_0'(0) = 0$ and $f_k(0) = 0$, $f_k'(0) = 0$ in case of $k \neq 0$. This means that all the conditions of Theorem 4.5 are satisfied and $f_0, f_1, \ldots, f_N$ form a generalized moment sequence.

To prove the converse we assume that the functions $f_0, f_1, \ldots, f_N$ form a generalized moment sequence and we prove by induction. It is obvious that the statement is true for f_0 and we suppose that $f_j(x) = \partial_t^j \varphi\big(x, f(t)\big)\big|_{t=0}$ for $j = 0, 1, \ldots, k$, where k is in $\{1, 2, \ldots, N\}$. We consider the function

$$g(x) = f_{k+1}(x) - \partial_t^{k+1} \varphi\big(x, f(t)\big)\big|_{t=0}$$

for each positive x. Then the expression $g''(x) + \frac{A'(x)}{A(x)} g'(x)$ is equal to

$$f''_{k+1}(x) + \frac{A'(x)}{A(x)} f'_{k+1}(x) - \partial_x^2 \partial_t^{k+1} \varphi(x, f(t))\big|_{t=0} - \frac{A'(x)}{A(x)} \partial_x \partial_t^{k+1} \varphi(x, f(t))\big|_{t=0}$$

and using Theorem 4.5 and equation (4.11) we get

$$c_0 f_{k+1}(x) + \sum_{j=1}^{k+1} \binom{k+1}{j} c_j \partial_t^{k+1-j} \varphi(x, f(t))\big|_{t=0}$$

$$- \sum_{j=0}^{k+1} \binom{k+1}{j} c_j \partial_t^{k+1-j} \varphi(x, f(t))|_{t=0} = c_0 \, g(x) \,.$$

Consequently

$$g''(x) + \frac{A'(x)}{A(x)} g'(x) = c_0 \, g(x) \,, \quad g(0) = 0, \quad g'(0) = 0 \,,$$

hence $g(x) \equiv 0$ and the proof is complete. $\qquad\square$

By this theorem a moment function of order N on K is a linear combination of functions of the form $x \mapsto \partial_2^j \varphi(x, \lambda)$, where $j = 1, 2, \ldots, N$ and λ is a complex number. If the coefficient of $\partial_2^N \varphi(x, \lambda)$ in this linear combination is different from zero, then we call the moment function *nondegenerate*, otherwise it is called *degenerate*.

Chapter 5

Two-point support hypergroups

5.1 Conditional functional equations

We shall use the following lemmas about some conditional functional equations in the sequel. These functional equations, called *cosine equation* (5.1) and *square-norm equation* (5.13), have been studied by several authors (see e.g. [Acz66], [Hos69], [CKN85] [Cor77], [Ste96], [FRS92]). Nevertheless, here we consider conditional versions of these functional equations in order to apply the result for the description of exponential and additive functions on some special hypergroups. For the solution of these conditional equations we shall use basic results on regularity in [Jár86], [JS96], [Jár05].

Lemma 1. *Let* $f : [0,1] \to \mathbb{C}$ *be a continuous function satisfying* $f(0) = 1$ *and*

$$f(x+y) + f(x-y) = 2f(x)f(y),$$ (5.1)

whenever $0 \le y \le x$ *and* $x + y \le 1$. *Then there exists a complex number* λ *such that* f *has the form*

$$f(x) = \cosh \lambda x$$ (5.2)

for each $x \ge 0$.

Proof. Let $\varphi = \operatorname{Re} f$ and $\psi = \operatorname{Im} f$, then $\varphi, \psi : [0,1] \to \mathbb{R}$ are continuous functions and we have

$$\varphi(x+y) + \varphi(x-y) = 2\varphi(x)\varphi(y) - 2\psi(x)\psi(y)$$ (5.3)

$$\psi(x+y) + \psi(x-y) = 2\varphi(x)\psi(y) + 2\psi(x)\varphi(y),$$ (5.4)

whenever $0 \le y \le x$ and $x + y \le 1$, further $\varphi(0) = 1$ and $\psi(0) = 0$. Let $0 < a \le 1$ such that $\varphi(x) > 0$ for $0 \le x \le a$.

Suppose first that φ and ψ are linearly dependent, that is, $\psi = c\varphi$ holds on $[0, 1]$ for some real c. By $\varphi(0) = 1$ and $\psi(0) = 0$ it follows $c = 0$, hence $\psi = 0$, which implies that $f = \varphi$ is real valued. If we let $A = \frac{a}{4}, B = \frac{a}{2}, C=0, D = \frac{a}{4}$, then we can apply Remark 22.12 in [Jár05] for the functional equation (5.1) on the intervals $]A, B[$ and $]C, D[$ to infer that $f = \varphi$ is C^∞ on the interval $]0, \frac{a}{4}[$.

If φ and ψ are linearly independent, then with the same choice of A, B, C, D we can apply the same result as above for the functional equation (5.3) on the intervals $]A, B[$ and $]C, D[$ to infer that φ and ψ are C^∞ on the interval $]0, \frac{a}{4}[$.

It follows that in any case f is C^∞ on some interval $]0, K[$, where $0 < K < 1$.

Let $m = \min\left\{\frac{5K}{4}, 1\right\}$. Suppose that $\frac{3K}{4} < t < m$, then, by (5.1), the substitution $x = t - \frac{K}{4}, y = \frac{K}{4}$ gives $0 \le y \le x \le 1$, $0 \le x + y \le 1$ and

$$f(t) = 2f\left(t - \frac{K}{4}\right) f\left(\frac{K}{4}\right) - f\left(t - \frac{K}{2}\right). \tag{5.5}$$

As $t - \frac{K}{4}$ and $t - \frac{K}{2}$ is in $]0, K[$, the right hand side is C^∞ on $]\frac{3K}{4}, m[$. It follows that f is C^∞ on $]0, m[$. If $\frac{5K}{4} \ge 1$, then we have that f is C^∞ on $]0, 1[$. If $\frac{5K}{4} < 1$, then replacing K by $\frac{5K}{4}$ and repeating the above argument after some steps we get that f is C^∞ on $]0, 1[$.

Differentiating (5.1) twice with respect to y and then substituting $y = 0$ we obtain that

$$f''(x) = cf(x) \tag{5.6}$$

for each x in $]0, 1[$ with $c = f''(0)$. As $f(0) = 1$, our statement follows. $\square$

Lemma 2. *Let* $f : [0, +\infty[\to \mathbb{C}$ *be a continuous function satisfying* $f(0) = 1$ *and*

$$f(x + y) + f(x - y) = 2f(x)f(y) \,, \tag{5.7}$$

whenever $0 \le y \le x$. *Then there exists a complex number* λ *such that* f *has the form*

$$f(x) = \cosh \lambda x \tag{5.8}$$

for each $x \ge 0$.

Proof. The proof is similar to that of the previous lemma. □

The following corollaries are easy consequences.

Corollary 1. *Let* $f : [0, 1] \to \mathbb{R}$ *be a continuous function satisfying* $f(0) = 1$ *and*

$$f(x + y) + f(x - y) = 2f(x)f(y) \,, \tag{5.9}$$

whenever $0 \le y \le x$ *and* $x + y \le 1$. *Then there exists a real number* λ *such that* f *has the form*

$$f(x) = \cosh \lambda x \quad or \quad f(x) = \cos \lambda x \tag{5.10}$$

for each $x \ge 0$.

Corollary 2. *Let* $f : [0, +\infty[\to \mathbb{R}$ *be a continuous function satisfying* $f(0) = 1$ *and*

$$f(x + y) + f(x - y) = 2f(x)f(y) \,, \tag{5.11}$$

whenever $0 \le y \le x$. *Then there exists a real number* λ *such that* f *has the form*

$$f(x) = \cosh \lambda x \quad or \quad f(x) = \cos \lambda x \tag{5.12}$$

for each $x \ge 0$.

Lemma 3. *Let* $f : [0, 1] \to \mathbb{C}$ *be a continuous function satisfying*

$$f(x + y) + f(x - y) = 2f(x) + 2f(y) \,, \tag{5.13}$$

whenever $0 \le y \le x$ *and* $x + y \le 1$. *Then there exists a complex number* λ *such that* f *has the form*

$$f(x) = \lambda x^2 \tag{5.14}$$

for each $x \ge 0$. *Moreover,* f *is real if and only if* λ *is real.*

Proof. Clearly $f(0) = 0$. Let $\varphi = Re\, f$ and $\psi = Im\, f$, then obviously $\varphi, \psi : [0, 1] \to \mathbb{R}$ are continuous functions and we have

$$\varphi(x + y) + \varphi(x - y) = 2\varphi(x) + 2\varphi(y) \qquad (5.15)$$

$$\psi(x + y) + \psi(x - y) = 2\psi(x) + 2\psi(y), \qquad (5.16)$$

whenever $0 \leq y \leq x$ and $x + y \leq 1$, further $\varphi(0) = \psi(0) = 0$. This means that the real and imaginary parts of f satisfy the same functional equation (5.13), hence we may suppose that f itself is real valued.

If f and 1 are linearly dependent, then f is constant: $f = 0$, hence our statement follows.

However, if f and 1 are linearly independent, then with the same choice of A, B, C, D applying the same result as above for the functional equation (5.13) on the intervals $]A, B[$ and $]C, D[$ to infer that f is C^∞ on some interval $]0, K[$, where $0 < K < 1$. Then, using the same argument as above, we infer that f is C^∞ on $]0, 1[$.

Differentiating (5.13) twice with respect to y, then substituting $y = 0$ and differentiating again we obtain that

$$f'''(x) = 0 \qquad (5.17)$$

for each x in $]0, 1[$, which implies that f is a quadratic polynomial on $[0, 1]$. Substituting into (5.13) our statement follows. The last assertion is obvious. $\qquad \square$

Lemma 4. *Let $f : [0, +\infty[\to \mathbb{C}$ be a continuous function satisfying*

$$f(x + y) + f(x - y) = 2f(x) + 2f(y), \qquad (5.18)$$

whenever $0 \leq y \leq x$. Then there exists a complex number λ such that f has the form

$$f(x) = \lambda x^2 \qquad (5.19)$$

for each $x \geq 0$. Moreover, f is real if and only if λ is real.

Proof. The proof is similar to that of the previous lemma. $\qquad \square$

5.2   Two-point support hypergroups of noncompact type

In this section we characterize exponential and additive functions on a two-point hypergroup of noncompact type. We use the lemmas proved in the previous section.

In [BH95] on p. 191 in Example 3.4.5 the authors present the following noncompact *two-point support* hypergroup. Let $K = [0, +\infty[$ and let the convolution of the Dirac-measures δ_x and δ_y be defined as

$$\delta_x * \delta_y = \frac{1}{2}(\delta_{x+y} + \delta_{|x-y|}) \tag{5.20}$$

for each $0 \leq x, y$. The name is given due to the fact that the support of $\delta_x * \delta_y$ consists of two points. We describe the additive and exponential functions on this hypergroup and give an exponential family (see Section 1.6) on it.

Theorem 5.1. *Let $(K, *)$ be the hypergroup defined as above. Then any exponential on K has the form*

$$m(x) = \cosh \lambda x\,, \tag{5.21}$$

whenever $x \geq 0$, where λ is an arbitrary complex number. Conversely, for each complex λ the function given in (5.21) is an exponential on K. Consequently, the function

$$(x, \lambda) \mapsto \cosh \lambda x$$

is an exponential family on K.

Proof. By the above definition of the convolution, the continuous function $m : [0, +\infty[\to \mathbb{C}$ is an exponential on K if and only if $m(0) = 1$ and the functional equation

$$m(x + y) + m(|x - y|) = 2m(x)m(y)$$

holds for each $x, y \geq 0$. This is equivalent to the condition

$$m(x + y) + m(x - y) = 2m(x)m(y)$$

for each $x \geq y \geq 0$. As obviously $m(0) = 1$, the statement follows from Lemma 2. □

From this result we get immediately the real exponentials.

Corollary 3. *Let $(K, *)$ be the hypergroup defined as above. Then any real exponential on K has the form*

$$m(x) = \cosh \lambda x, \text{ or } m(x) = \cos \lambda x, \tag{5.22}$$

whenever $x \geq 0$, where λ is an arbitrary real number. Conversely, for each real λ the function given in (5.22) is a real exponential on K.

We recall from Section 1.5 that bounded real exponentials are exactly the characters, which, by (5.22), have the form $x \mapsto \cos \lambda x$ with some real λ.

Now we turn to additive functions on the hypergroup K.

Theorem 5.2. *Let $(K, *)$ be the hypergroup defined as above. Then any additive function on K has the form*

$$a(x) = \lambda x^2, \tag{5.23}$$

whenever $x \geq 0$, where λ is an arbitrary complex number.

Proof. By the above definition of the convolution, the continuous function $a : [0, +\infty[\to \mathbb{C}$ is an additive function on K if and only if the functional equation

$$a(x + y) + a(|x - y|) = 2a(x) + 2a(y)$$

holds for each $x, y \geq 0$. This is equivalent to the condition

$$a(x + y) + a(x - y) = 2a(x) + 2a(y)$$

for each $x \geq y \geq 0$. Our statement follows from Lemma 4. □

5.3   Moment functions on two-point support hypergroups of noncompact type

This section is devoted to the characterization of moment functions on the two-point hypergroup $K = [0, +\infty[$ in the previous section. We recall (see Section 1.8) that the sequence of continuous functions $\varphi_n : K \to \mathbb{C}$ ($n = 0, 1, \dots$) is a moment function sequence, if the system of functional equations

$$\varphi_n(x * y) = \sum_{k=0}^{n} \binom{n}{k} \varphi_k(x) \varphi_{n-k}(y) \qquad (5.24)$$

holds for each x, y in K and for $n = 0, 1, \ldots$. In our case this means that we have

$$\varphi_n(x + y) + \varphi_n(|x - y|) = 2 \sum_{k=0}^{n} \binom{n}{k} \varphi_k(x) \varphi_{n-k}(y) \qquad (5.25)$$

for each x, y in K and for $n = 0, 1, \ldots$, that is,

$$\varphi_n(x + y) + \varphi_n(x - y) = 2 \sum_{k=0}^{n} \binom{n}{k} \varphi_k(x) \varphi_{n-k}(y) \qquad (5.26)$$

for each $0 \le y \le x$ and for $n = 0, 1, \ldots$. For this system of functional equations we can apply the results in [Jár05] again, using the same arguments as above. We get the following result.

Theorem 5.3. *Let $(\varphi_n)_{n \in \mathbb{N}}$ be a generalized moment function sequence on the hypergroup K. Then φ_n is C^∞ for each n in $\mathbb{N}$, further $\varphi_n(0) = 0$ for $n = 1, 2, \ldots$ and $\varphi_n'(0) = 0$ for $n = 0, 1, \ldots$.*

Proof. Exactly in the same way as above, using the results in [Jár05] we infer from (5.25) that φ_n is C^∞ on $[0, +\infty[$. We note that differentiability of any order at 0 means obviously right differentiability and the corresponding derivatives are right derivatives.

As φ_0 is an exponential we have that $\varphi_0(0) = 1$. For $n = 1$ we have

$$\varphi_1(x * y) = \varphi_1(x)\varphi_0(y) + \varphi_0(x)\varphi_1(y)$$

for each $x, y \ge 0$. Substituting $y = 0$ it follows

$$\varphi_1(x) = \varphi_1(x) + \varphi_1(0)\varphi_0(x),$$

hence $\varphi_1(0) = 0$. Suppose that $n \ge 2$ and we have proved $\varphi_k(0) = 0$ for $k = 1, 2, \ldots, n - 1$. Then we obtain

$$\varphi_n(x * y) = \sum_{k=1}^{n-1} \binom{n}{k} \varphi_k(x)\varphi_{n-k}(y) + \varphi_0(x)\varphi_n(y) + \varphi_n(x)\varphi_0(y),$$

for each $x, y \geq 0$. Putting $y = 0$ it follows

$$\varphi_n(x) = \varphi_0(x)\varphi_n(0) + \varphi_n(x),$$

which implies $\varphi_n(0) = 0$. We note that this part of the proof works on any hypergroup.

For the proof of the second statement we use equation (5.26). For $n = 0$ we obtain

$$\varphi_0(x + y) + \varphi_0(x - y) = 2\varphi_0(x)\varphi_0(y)$$

for each $0 \leq y \leq x$. Differentiating this equation with respect to y we get

$$\varphi_0'(x + y) + \varphi_0'(x - y) = 2\varphi_0(x)\varphi_0'(y)$$

for each $0 \leq y \leq x$. Putting $y = 0$ we have

$$\varphi_0(x)\varphi_0'(0) = 0$$

for each $x \geq 0$, hence $\varphi_0'(0) = 0$. Let $n \geq 0$ be given and suppose that we have proved $\varphi_k'(0) = 0$ for $k = 0, 1, \ldots, n - 1$ ($n \geq 1$). We differentiate equation (5.26) with respect to y, then we get

$$\varphi_n'(x + y) - \varphi_n'(x - y) = 2\sum_{k=0}^{n} \binom{n}{k} \varphi_k(x)\varphi_{n-k}'(y) \qquad (5.27)$$

for each $0 \leq y \leq x$. Putting $y = 0$ it follows

$$\varphi_0(x)\varphi_n'(0) = 0$$

for each $x \geq 0$, hence $\varphi_n'(0) = 0$. The theorem is proved. $\square$

Now we are ready to characterize generalized moment functions on the hypergroup K. We have to consider two cases. The first one is the case of the ordinary moment function sequences, that is, the case, where the starting exponential is identically 1. It turns out that in this case the sequence of moment functions consists of polynomials and they can be determined recursively from a simple system of linear differential equations of second order. The corresponding result follows.

Theorem 5.4. *Let n be a nonnegative integer and let $\varphi_k : [0, +\infty[\to \mathbb{C}$ ($k = 0, 1, \ldots, n$) be continuous functions with $\varphi_0 = 1$, which form a moment function sequence on the hypergroup K, that is,*

$$\varphi_k(x * y) = \sum_{j=0}^{k} \binom{k}{j} \varphi_j(x)\varphi_{k-j}(y) \tag{5.28}$$

holds for each $x, y \geq 0$ and $k = 0, 1, \ldots, n$. Then φ_k is an even polynomial of degree at most $2k$ for $k = 1, 2, \ldots, n$ satisfying

$$\varphi_k''(x) = \sum_{j=0}^{k-1} \binom{k}{j} \varphi_j(x)\varphi_{k-j}''(0) \tag{5.29}$$

for each $x \geq 0$ and

$$\varphi_k(0) = \varphi_k'(0) = 0 \tag{5.30}$$

for $k = 1, 2, \ldots, n$.

Proof. By the definition of the convolution in K we have

$$\varphi_k(x + y) + \varphi_k(x - y) = 2\sum_{j=0}^{k} \binom{k}{j} \varphi_j(x)\varphi_{k-j}(y) \tag{5.31}$$

for each $0 \leq y \leq x$ and $k = 0, 1, \ldots, n$ further $\varphi_0 = 1$. Differentiating this equation twice with respect to y and then substituting $y = 0$ we obtain equation (5.29). The initial conditions (5.30) are satisfied for any moment functions of degree at least 2. For $k = 1$ we get $\varphi_1(x) = cx^2$ with some complex number c. By induction it follows that the right hand side of (5.29) is an even polynomal of degree at most $2(k - 1)$, hence the solution satisfying (5.30) is a polynomial of degree at most $2k$ and our theorem is proved. $\square$

In the second case, where the starting exponential in the sequence of generalized moment functions is nonconstant our point is to show that these generalized moment functions can be obtained from the exponential family exactly in the same manner as in the case of polynomial hypergroups. We shall see in the following section that this general idea works on another important type of hypergroups: on the Sturm–Liouville hypergroups.

Let λ be a complex number, n a nonnegative integer, $c_0 = \lambda$ and $c_1, c_2, \ldots, c_n$ arbitrary complex numbers. We define the function $f : \mathbb{R} \to \mathbb{C}$ by

$$f(t) = \sum_{k=0}^{n} c_k \frac{t^k}{k!} \tag{5.32}$$

and the sequence

$$\psi_k(x) = \frac{d^k}{dt^k} \big(\cosh f(t)x \big)\big|_{t=0} \tag{5.33}$$

for each t in $\mathbb{R}$ and $x \geq 0$. The complex numbers $c_0, c_1, \ldots, c_n$ can be considered as parameters in the definition of ψ_k but – in order to avoid complicated notation – we will not explicitly denote this dependence of ψ_k on these "hidden" variables. In particular, $\psi_0(x) = \cosh f(0)x = \cosh \lambda x$ holds for $x \geq 0$.

Our first theorem shows that the functions $\psi_0, \psi_1, \ldots, \psi_n$ form a generalized moment sequence on K for any choice of the parameters.

Theorem 5.5. *Let the complex numbers $c_0, c_1, \ldots, c_n$ be arbitrary. Then the functions $\psi_0, \psi_1, \ldots, \psi_n$ form a generalized moment sequence, that is,*

$$\psi_k(x * y) = \sum_{j=0}^{n} \binom{k}{j} \psi_j(x) \psi_{k-j}(y) \tag{5.34}$$

holds for $k = 0, 1, \ldots, n$ and for each $x, y \geq 0$.

Proof. Clearly we have

$$\cosh f(t)(x+y) + \cosh f(t)(x-y) = 2 \cosh f(t)x \cosh f(t)y \tag{5.35}$$

for each t, x, y in $\mathbb{R}$. Using Leibniz Rule and differentiating this equation k times with respect to t, then substituting $t = 0$ we get for every $x, y \geq 0$

$$\psi_k(x * y) = \sum_{j=0}^{k} \binom{k}{j} \frac{d^j}{dt^j} \left. (\cosh f(t)x) \right|_{t=0} \frac{d^{k-j}}{dt^{k-j}} \left. (\cosh f(t)y) \right|_{t=0}$$

$$= \sum_{j=0}^{k} \binom{k}{j} \psi_j(x) \psi_{k-j}(y).$$

Our statement is proved. □

The next theorem shows that the family of sequences $\psi_0, \psi_1, \ldots, \psi_n$ corresponding to all possible choices of the parameters $c_0, c_1, \ldots, c_n$ include all generalized moment function sequences with $\varphi_0 \neq 1$ on K: every generalized moment function sequence with $\varphi_0 \neq 1$ on K can be obtained from this family by an appropriate choice of the parameters.

Theorem 5.6. *Let n be a nonnegative integer and let $\varphi_k : [0, +\infty[\to \mathbb{C}$ ($k = 0, 1, \ldots, n$) be continuous functions with $\varphi_0 \neq 1$, which form a generalized moment function sequence on the hypergroup K, that is,*

$$\varphi_k(x * y) = \sum_{j=0}^{k} \binom{k}{j} \varphi_j(x) \varphi_{k-j}(y) \tag{5.36}$$

holds for each $x, y \geq 0$ and $k = 0, 1, \ldots, n$. Then there exist complex numbers $c_0, c_1, \ldots, c_n$ such that for the corresponding functions ψ_k built up in (5.32) and (5.33) on these parameters we have

$$\varphi_k = \psi_k \tag{5.37}$$

holds for $k = 0, 1, \ldots, n$.

Proof. For $k = 0$ we have from (5.36) that φ_0 is an exponential on K and Theorem 5.1 implies that there exists a complex number λ such that

$$\varphi_0(x) = \cosh \lambda x$$

holds for all x in K. Let $c_0 = \lambda$. By our assumption $\lambda \neq 0$. Then $\varphi_0 = \psi_0$, in particular $\varphi_0''(0) = \lambda^2 = \psi_0''(0)$.

By Theorem 5.3 φ_k is C^∞ for $k = 0, 1, \ldots, n$ and we have

$$\varphi_k(x+y) + \varphi_k(x-y) = 2\sum_{j=0}^{k} \binom{k}{j} \varphi_j(x)\varphi_{k-j}(y) \qquad (5.38)$$

holds for each $0 \le y \le x$ and $k = 0, 1, \ldots, n$. Let $k \ge 1$ be fixed. We differentiate (5.38) two times with respect to y and substitute $y = 0$ we get that

$$\varphi_k''(x) - \lambda^2 \varphi_k(x) = \sum_{j=0}^{k-1} \binom{k}{j} \varphi_j(x)\varphi_{k-j}''(0) \qquad (5.39)$$

holds for each $0 \le x$ and $k = 1, \ldots, n$. Suppose that we have chosen the complex numbers $c_0, c_1, \ldots, c_{k-1}$ such that $\varphi_j = \psi_j$ holds, whenever $j = 0, 1, \ldots, c_{j-1}$ and we want to make a choice for c_k such that $\varphi_k = \psi_k$ holds. According to our assumption we can rewrite equation (5.39) as

$$\varphi_k''(x) - \lambda^2 \varphi_k(x) = \sum_{j=1}^{k-1} \binom{k}{j} \psi_j(x)\psi_{k-j}''(0) + \varphi_k''(0)\psi_0(x) \qquad (5.40)$$

holds for each $0 \le x$. Further, by Theorem 5.3, we have

$$\varphi_k(0) = \varphi_k'(0) = 0. \qquad (5.41)$$

This means that φ_k is the solution of the second order linear differential equation (5.40) with the initial data (5.41). By the uniqueness of the solution of this Cauchy-problem it is enough to show that we can choose a complex number c_k in the way that the corresponding ψ_k is a solution of the same Cauchy-problem, too. As equation (5.40) has been derived from equation (5.38) and, by Theorem 5.34, the functions ψ_j also satisfy (5.38), it follows that ψ_k definitely will satisfy (5.40) if we can choose c_k in such a way that $\psi_k''(0) = \varphi_k''(0)$, beacause in this case the right hand side of equation (5.40) for φ_k and for ψ_k will be identical. Moreover, we have seen in Theorem 5.3 that ψ_k satisfies the initial conditions (5.41).

Unfortunately, from the explicit form of ψ_k it is not clear the dependence of $\psi_k''(0)$ on the parameter c_k. On the other hand, we actually do not need to know much about this dependence: it is enough to show that $\psi_k''(0)$ properly depends on c_k. By the Implicit Function Theorem it is enough to show that the derivative of the function $c_k \mapsto \psi_k''(0)$ is nonzero. By the definition of ψ_k we will show that the following derivative is nonzero:

$$\frac{d}{dc_k} \frac{d^2}{dx^2} \frac{d^k}{dt^k} (\cosh f(t)x) \Big|_{t=0, x=0} \tag{5.42}$$

is nonzero. It turns out that this derivative is $2\lambda \neq 0$, independently of the values of $c_1, c_2, \ldots, c_{k-1}$. Indeed, using the fact that, by definition, the function

$$(t, x, c_k) \mapsto \left(\cosh \sum_{j=0}^{k} c_j \frac{t^j}{j!} x \right)$$

is clearly $\mathcal{C}^\infty$, the order of differentiation can be interchanged and we easily get our statement on the value of the derivative in (5.42): it is 2λ, consequently, by the Implicit Function Theorem, c_k can be expressed as a (unique) function of $\psi_k''(0)$ and our theorem is proved. $\square$

We can summarize our results in the following theorem, which contains the complete description of generalized moment function sequences on the two-point support hypergroup of noncompact type K.

Theorem 5.7. *Let n be a nonnegative integer and let $\varphi_k : [0, +\infty[\to \mathbb{C}$ $(k = 0, 1, \ldots, n)$ be continuous functions, which form a generalized moment function sequence on the hypergroup K. Then we have the following possibilities:*

(1) If $\varphi_0 = 1$, then φ_k has the form

$$\varphi_k(x) = \sum_{i=1}^{k} a_{k,i} x^{2i}$$

for all $x \geq 0$, where $a_{k,i}$ $(k = 1, 2, \ldots, n, i = 1, 2, \ldots, k)$ are complex numbers chosen in the manner that the polynomials $\varphi_0, \varphi_1, \ldots, \varphi_n$ satisfy the initial value problem:

$$\varphi_k''(x) = \sum_{j=0}^{k-1} \binom{k}{j} \varphi_j(x) \varphi_{k-j}''(0) \tag{5.43}$$

for each $x \geq 0$ and

$$\varphi_k(0) = \varphi_k'(0) = 0 \tag{5.44}$$

for $k = 1, 2, \ldots, n$.

(2) If $\varphi_0 \neq 1$, then there exist complex numbers $c_0, c_1, \ldots, c_n$ such that

$$\varphi_k(x) = \frac{d^k}{dt^k} \left(\cosh f(t) x \right) \big|_{t=0} \tag{5.45}$$

holds for $x \geq 0$ and $k = 0, 1, \ldots, n$, where

$$f(t) = \sum_{j=0}^{n} \frac{c_j}{j!} t^j$$

for each t in $\mathbb{R}$.

5.4   Two-point support hypergroups of compact type

In this section we characterize exponential and additive functions on a two-point hypergroup of compact type. We use again the lemmas proved in the previous section.

In [BH95] on p. 191, Example 3.4.5 is the following compact two-point support hypergroup. Let $K = [0, 1]$ and let the convolution of the Dirac-measures δ_x and δ_y be defined as

$$\delta_x * \delta_y = \frac{1}{2} \left(\delta_{|x-y|} + \delta_{|1-|1-x-y||} \right) \tag{5.46}$$

for each $0 \leq x, y \leq 1$.

Theorem 5.8. *Let $(K, *)$ be the hypergroup defined as above. Then any exponential on K has the form*

$$m(x) = \cosh \lambda x, \tag{5.47}$$

whenever $0 \leq x \leq 1$, where λ is an arbitrary complex number. Conversely, for each complex λ the function given in (5.47) is an exponential on K. Consequently, the function

$$(x, \lambda) \mapsto \cosh \lambda x$$

is an exponential family on K.

Proof.   Similarly, as in the proof of Theorem 5.1, by the definition of the convolution, the continuous function $m : [0, 1] \to \mathbb{C}$ is an exponential on K if and only if $m(0) = 1$ and the functional equation

$$m(x + y) + m(|x - y|) = 2m(x)m(y)$$

holds for each $0 \leq x, y \leq 1$. This is equivalent to the condition

$$m(x + y) + m(x - y) = 2m(x)m(y)$$

for each $0 \leq y \leq x \leq 1$ with $x + y \leq 1$. The statement follows from Lemma 1. $\qquad\square$

Again, we have that real exponentials have the form $x \mapsto \cosh \lambda x$ and $x \mapsto \cos \lambda x$ with real λ, while the general form of characters is $x \mapsto \cos \lambda x$ with real λ.

The description of additive functions is equally simple.

Theorem 5.9. *Let $(K, *)$ be the hypergroup defined as above. Then any additive function on K has the form*

$$a(x) = \lambda x^2 \,, \qquad\qquad (5.48)$$

whenever $0 \leq x \leq 1$, where λ is an arbitrary complex number.

Proof. We use Lemma 3. $\qquad\square$

The problem of determination of generalized moment function sequences on this two-point support hypergroup of compact type can be settled along exactly the same lines like in Section 5.3. Namely, we can reduce the solution of the system of equations characterizing generalized moment function sequences to the solution of similar differential equations like above and we obtain the same solutions. The straightforward results are summarized in Theorem 5.7, the interval $[0, +\infty[$ replaced by $[0, 1]$. The details are left to the reader.

5.5 The cosh hypergroup

Our third special hypergroup in this section is a *cosh hypergroup* studied in some details in [Zeu89] (see also [BH95], p. 191, Example 3.4.6). Here $K = [0, +\infty[$ and the convolution is defined by

$$\delta_x * \delta_y = \frac{\cosh(x + y)}{2 \cosh x \cosh y} \delta_{x+y} + \frac{\cosh(|x - y|)}{2 \cosh x \cosh y} \delta_{|x-y|} \qquad (5.49)$$

for each $x, y \geq 0$.

Theorem 5.10. *Let* $(K, *)$ *be the hypergroup defined as above. Then any exponential on* K *has the form*

$$m(x) = \frac{\cosh \lambda x}{\cosh x}, \qquad (5.50)$$

whenever $x \geq 0$, *where* λ *is an arbitrary complex number. Conversely, for each complex* λ *the function given in* (5.50) *is an exponential on* K. *Consequently, the function*

$$(x, \lambda) \mapsto \frac{\cosh \lambda x}{\cosh x}$$

is an exponential family on K.

Proof. By the above definition of convolution on K the continuous function $m : [0, +\infty[\to \mathbb{C}$ is an exponential on K if and only if $m(0) = 1$ and it satisfies the functional equation

$$\frac{\cosh(x+y)}{\cosh x \cosh y} m(x+y) + \frac{\cosh(x-y)}{\cosh x \cosh y} m(x-y) = 2m(x)m(y) \qquad (5.51)$$

for each $0 \leq y \leq x$. Multiplying by $\cosh x \cosh y$ we have that the continuous function

$$\varphi(x) = \cosh x \, m(x)$$

satisfies $\varphi(0) = 1$ and the equation (5.7) for $0 \leq y \leq x$. It follows that φ has the form given in (5.8) and our statements follow. $\qquad \square$

5.6   Associated pairs of moment functions

In this section we show how to determine associated pairs on moment functions using our above results on the two-point support hypergroups of compact and noncompact type considered in Sections 5.3 and 5.4. Concerning associated pairs of moment and generalized moment functions see also Section 1.8 and the works [Gal98], [GT02b], [Gal97], [BH95].

First we remark that if K denotes any of the two hypergroups studied in Sections 5.2 and 5.4, then all generalized moment sequences have been described in Theorem 5.7 and they have the same form in both cases. This means that in the remaining part of this section K denotes either of the two hypergroups studied in Sections 5.2 and 5.4. Suppose that the generalized associated pair of moment functions $\varphi_1, \varphi_2 : K \to \mathbb{C}$ is generated by the exponential function $\varphi_0 : K \to \mathbb{C}$. By the terminology introduced in Section 1.8 and by the above results this means that φ_0, φ_1 and φ_2 are real valued $\mathcal{C}^\infty$-functions, φ_2 is nonnegative, the inequality

$$\varphi_1^2(x) \le \varphi_0(x) \cdot \varphi_2(x) \qquad (5.52)$$

holds for each x in K and these three functions are the first three terms of a generalized moment function sequence described in Theorem 5.7. Corresponding to Corollary 3, λ is real and we shall separate three different cases. In the first case we suppose that in the representation (5.22) of the exponential φ_0 we have $\lambda = 0$, that is, $\varphi_0 = 1$ for each x in K. Actually, instead of using Theorem 5.7 here we can solve the corresponding differential equations easily. We have that

$$\varphi_1(x + y) + \varphi_1(x - y) = 2\varphi_1(x) + 2\varphi_1(y) \qquad (5.53)$$

holds for each $0 \le y \le x$ in K. By Lemmas 4 and 3 in Section 5.1 it follows that $\varphi_1 = ax^2$ for all x in K, where a is a real number. Further we have

$$\varphi_2(x + y) + \varphi_2(x - y) = 2\varphi_2(x) + 4a^2x^2y^2 + 2\varphi_2(y) \qquad (5.54)$$

holds for each x, y in K, if $x - y$ is in K, too. We differentiate (5.54) twice with respect to y and substitute $y = 0$ to obtain the differential equation

$$\varphi_2''(x) = \varphi_2(0) + 4a^2x^2$$

for x in K. The solution of this differential equation is

$$\varphi_2(x) = \frac{1}{3}a^2x^4 + bx^2 \qquad (5.55)$$

for each x in K. By $\varphi_2 \geq 0$ we have $b = c^2 \geq 0$ and the inequality (5.52) implies that $a = 0$. Hence in this case we have the trivial associated pair of moment functions $0, c^2 x^2$ generated by he exponential 1.

Now we turn to the case where $\lambda \neq 0$ is a real number and $\varphi_0(x) = \cosh \lambda x$ for each x in K. We can apply Theorem 5.7 to get

$$\varphi_1(x) = c_1 x \sinh \lambda x \,,$$

$$\varphi_2 = c_1^2 x^2 \cosh \lambda x + c_2 x \sinh \lambda x$$

for all x in K, where c_1, c_2 are real numbers. The condition $\varphi_2 \geq 0$ means that

$$c_1^2 x^2 \cosh \lambda x + c_2 x \sinh \lambda x \geq 0 \,,$$

that is

$$c_1^2 \cosh \lambda x + c_2 \lambda \frac{\sinh \lambda x}{\lambda x} \geq 0$$

holds for each $x \neq 0$ in K. Letting $x \to 0$ we get that

$$c_1^2 + c_2 \lambda \geq 0 \tag{5.56}$$

is a necessary condition for φ_1, φ_2 to form a generalized associated pair generated by φ_0. To prove its sufficiency we consider the function

$$F(x) = c_1^2 x \cosh \lambda x + c_2 \sinh \lambda x$$

for x in K. The derivative

$$F'(x) = (c_1^2 + c_2 \lambda) \cosh \lambda x + \lambda^2 c_1^2 x^2 \frac{\sinh \lambda x}{\lambda x} \,,$$

which is positive for all x in K, assuming (5.56). Hence F is strictly increasing. By $F(0) = 0$ this means that if (5.56) holds, then F, hence also φ_2 is nonnegative.

On the other hand, we have for $x > 0$ in K

$$\varphi_0(x) \cdot \varphi_2(x) - \varphi_1^2(x) = c_2 x \sinh \lambda x \cosh \lambda x + c_1^2 x^2 (\cosh^2 \lambda x - \sinh^2 \lambda x)$$

$$= c_2 \lambda x^2 \frac{\sinh 2\lambda x}{2\lambda x} + c_1^2 x^2 \cosh 2\lambda x \geq x^2 (c_2 \lambda + c_1^2)(\frac{\sinh 2\lambda x}{2\lambda x} + \cosh \lambda x) \geq 0 \,,$$

whenever (5.56) holds, which shows that this condition is necessary and sufficient for φ_1, φ_2 to form a generalized associated pair generated by φ_0.

Hence we have proved our first result on generalized associated pairs generated by nonconstant exponentials on K.

Theorem 5.11. *Let K be either of the two two-point support hypergroups considered in Sections 5.4 and 5.2. The only generalized associated pair generated by a convex exponential is (φ_1, φ_2) generated by φ_0, where*

$$\varphi_0(x) = \cosh \lambda x \,,$$

$$\varphi_1(x) = c_1 x \sinh \lambda x$$

and

$$\varphi_2(x) = c_1^2 x^2 \cosh \lambda x + c_2 x \sinh \lambda x$$

for all x in K, where λ, c_1, c_2 are real numbers, λ is nonzero and c_1, c_2 satisfy

$$c_2 \lambda + c_1^2 \geq 0 \,.$$

Now we consider the case where $\lambda \neq 0$ is a real number and $\varphi_0(x) = \cos \lambda x$ for each x in K. We proceed similarly like above. Namely, we apply again Theorem 5.7 to get

$$\varphi_1(x) = -c_1 x \sin \lambda x \,, \tag{5.57}$$

$$\varphi_2 = -c_2 x \sin \lambda x - c_1^2 x^2 \cos \lambda x \qquad (5.58)$$

for all $x \neq 0$ in K, where c_1, c_2 are real numbers. The condition $\varphi_2 \geq 0$ means that

$$-c_2 x \sin \lambda x - c_1^2 x^2 \cos \lambda x \geq 0 \,,$$

that is

$$c_2 \lambda \frac{\sin \lambda x}{\lambda x} + c_1^2 \cos \lambda x \leq 0$$

for each $x \neq 0$ in K. For $x \to 0$ this implies

$$c_1^2 + c_2 \lambda \leq 0 \,, \qquad (5.59)$$

hence this is a necessary condition for φ_1, φ_2 to form a generalized associated pair generated by φ_0. We show that in the case of the noncompact hypergroup $K = [0, +\infty[$ this can be satisfied only in the trivial case $\varphi_1 = \varphi_2 = 0$. Indeed, if $\lambda > 0$ and we let $x = \frac{\pi}{2\lambda}$, then

$$\varphi_2(x) = -c_2 \frac{\pi}{2\lambda} \geq 0$$

implies that $c_2 \leq 0$. On the other hand, for $x = \frac{3\pi}{2\lambda}$ it follows

$$\varphi_2(x) = c_2 \frac{3\pi}{2\lambda} \geq 0 \,,$$

hence $c_2 \geq 0$, thus it follows $c_2 = 0$ and in this case, by (5.59), we have $c_1 = 0$. We have proved the following result:

Theorem 5.12. *Let K be the noncompact type two-point support hypergroup considered in Section 5.2. The only generalized associated pair generated by a nonconvex exponential is the trivial $(0,0)$.*

The only case left and we have to consider is the compact hypergroup K in Section 5.4 and the exponential $\varphi_0 : x \mapsto \cos \lambda x$ with a nonzero real λ, further φ_1, φ_2 of the form given in (5.57) and (5.58), where c_1, c_2 are real numbers satisfying (5.59). Then we have that a necessary and sufficient condition for φ_1, φ_2 form a generalized associated pair generated by φ_0 is that for each $0 \le x \le 1$ we have

$$\varphi_2(x) = -c_1^2 x^2 \cos \lambda x - c_2 x \sin \lambda x \ge 0 \tag{5.60}$$

and, by (5.52)

$$\varphi_2(x) \cdot \varphi_0(x) - \varphi_1(x)^2 = -c_2 x \sin \lambda x \cos \lambda x - c_1^2 x^2 \cos^2 \lambda x - c_1^2 x^2 \sin^2 \lambda x$$

$$= -\frac{1}{2} x \sin 2\lambda x - c_1^2 x^2 \ge 0 \,. \tag{5.61}$$

If $\lambda \ge 2\pi$, then $0 \le \frac{2\pi}{\lambda} \le 1$ and we let $x = \frac{2\pi}{\lambda}$. It follows by (5.60)

$$\varphi_2(x) = -c_1^2 \frac{4\pi^2}{\lambda^2} \ge 0 \,,$$

which implies $c_1 = 0$. Similarly, in the case $\lambda \le -2\pi$ we get the same conclusion. Further, if $\pi \le |\lambda| < 2\pi$, then $\frac{\pi}{|\lambda|} \le 1$ and $\pi \le |\lambda x| \le 2\pi$, which means that the sign of $\sin \lambda x$ is the opposite to the sign of λ, that is, the same as the sign of c_2, hence $\varphi_2(x) \le 0$ and, by (5.60), $\varphi_2 = 0$ follows.

Theorem 5.13. *Let K be the compact type two-point support hypergroup considered in Section 5.4. If $|\lambda| \ge \pi$, then the only generalized associated pair generated by a nonconvex exponential is the trivial $(0, 0)$.*

Chapter 6

Spectral analysis and synthesis on polynomial hypergroups

6.1 Spectral analysis and spectral synthesis on hypergroups

Spectral synthesis deals with the description of translation invariant function spaces over topological groups. Suppose that a locally compact Abelian group is given and consider the set of all continuous complex valued functions on it, equipped with the pointwise linear operations and with the topology of uniform convergence on compact sets. In order to set up the problem of spectral analysis and spectral synthesis in this context we have to define exponential functions and exponential monomials on commutative topological groups. Continuous homomorphisms of such groups into the additive topological group of complex numbers, and into the multiplicative topological group of nonzero complex numbers are called *additive functions*, and *exponential functions*, or *exponentials*, respectively. A *polynomial* on such a group is a polynomial of additive functions. An *exponential monomial* is a product of a polynomial and an exponential function. An *exponential polynomial* is a sum of exponential monomials. Now the problem of spectral analysis and spectral synthesis can be formulated: is it true that any nonzero closed translation invariant linear subspace of the space mentioned above (in other words a *variety*) contains an exponential function (*spectral analysis*), and is it true that in any variety the linear hull of all exponential monomials is dense (*spectral synthesis*)? The spectral analysis question can be reformulated in the following way: is it true that any variety contains a minimal (that means, one-dimensional) variety? A possible reformulation of the spectral synthesis problem asks if any variety is the sum of finite dimensional varieties. Concerning spectral analysis and spectral synthesis problems on locally compact Abelian groups the reader is referred to [Sch47], [Sch48], [Lef58],

[Szé91], [Ben75], [Beu48], [Kap49], [Dit39], [Ehr55a], [Ehr55b], [Hel83], [Ell65], [Lai72], [Gil66], [Hel52], [Mal59], [Szé02], [Szé06a], [Szé01]. In particular, spectral synthesis problems are studied in [Vog87], [CK79], [CR78], [KS81], too.

Here we formulate the basic problems of spectral analysis and spectral synthesis on commutative hypergroups and solve these problems on some types of them.

6.2   Basic concepts and facts

The presence of translation operators on hypergroups leads to the concept of variety. Let K be a locally compact Hausdorff space and let $\mathcal{C}(K)$ denote the locally convex topological vector space of all continuous complex valued functions on K, equipped with the pointwise linear operations and with the topology of uniform convergence on compact sets. The dual of this space can be identified with $\mathcal{M}_c(K)$, the latter being endowed with the weak*-topology with respect to the space of complex valued continuous functions on K. If, in addition, K is equipped with a commutative hypergroup structure, then a subset H of $\mathcal{C}(K)$ is called *translation invariant*, if for any f in H the function $\tau_y f$ belongs to H for all y in K. A nonzero closed translation invariant subspace of $\mathcal{C}(K)$ is called a *variety*. For any f in $\mathcal{C}(K)$ the *variety generated by* f is the closed subspace generated by all translates of f, which is denoted by $\tau(f)$.

The dual of $\mathcal{C}(K)$ is a locally convex topological vector space, which bears a natural algebra structure, corresponding to the convolution of measures. It is easy to see ([BH95]) that for any continuous function f in $\mathcal{C}(K)$ the function $(x, y) \mapsto f(x * y)$ is continuous. For any measures μ, ν in $\mathcal{M}_c(K)$ and for any f in $\mathcal{C}(K)$ we let

$$(\mu * \nu)(f) = \int_K \int_K f(x * y) \, d\mu(x) \, d\nu(y) \,.$$

Then $\mu * \nu$ is an element of $\mathcal{M}_c(K)$, which is called the *convolution* of μ and ν. The space $\mathcal{M}_c(K)$ equipped with the pointwise linear operations and with the convolution is a commutative algebra with unit.

For any closed linear subspace V in $\mathcal{C}(K)$ its *annihilator* $V^\perp$ in $\mathcal{M}_c(K)$ is the set of all measures from $\mathcal{M}_c(K)$ which vanish on V. Clearly, it is a closed linear subspace of $\mathcal{M}_c(K)$. The dual correspondence is also true: the annihilator $I^\perp$ of any closed linear subspace I of $\mathcal{M}_c(K)$, that is, the set of all elements of $\mathcal{C}(K)$, which belong to the kernel of all linear functionals in I is a closed linear subspace of $\mathcal{C}(K)$. By the Hahn–Banach Theorem we have the obvious relations $V = V^{\perp\perp}$ and $I = I^{\perp\perp}$ for any closed linear subspace V of $\mathcal{C}(K)$ and for any closed linear subspace I of $\mathcal{M}_c(K)$. In the case of varieties the annihilators can be characterized (see [Szé91]).

Theorem 6.1. *Let K be a commutative hypergroup, V a variety in $\mathcal{C}(K)$ and I a proper closed ideal in $\mathcal{M}(G)$. Then $V^\perp$ is a proper closed ideal in $\mathcal{M}(G)$ and $I^\perp$ is a variety in $\mathcal{C}(G)$.*

Proof. If f is an element of V, μ is an element of $V^\perp$ and ν is arbitrary in $\mathcal{M}_c(K)$, then by definition

$$(\mu * \nu)(f) = \int_K \int_K f(x * y)\, d\mu(x)\, d\nu(y) = \int_K \Big[\int_K T_y f(x)\, d\mu(x) \Big] d\nu(y) = 0\,,$$

hence $\mu * \nu$ belongs to $V^\perp$. As V is nonzero, hence $V^\perp$ is a proper ideal. Conversely, if $I = I^{\perp\perp}$ is a proper closed ideal in $\mathcal{M}_c(K)$, then for any μ in I, f in $I^\perp$ and ν in $\mathcal{M}_c(K)$ we have

$$0 = \int_K f\, d(\mu * \nu) = \int_K \Big[\int_K f(x * y)\, d\mu(x) \Big] d\nu(y)\,,$$

that is, the function $y \mapsto \mu(\tau_y f)$ annihilates $\mathcal{M}_c(K)$. This means that this function vanishes and, by definition, $\tau_y f$ belongs to $I^\perp$ for all y in K. $\square$

Let V be a variety in $\mathcal{C}(K)$. We say that *spectral analysis holds for V*, if V contains a one-dimensional (hence minimal) variety. If spectral analysis holds for any variety in $\mathcal{C}(K)$, then we say that *spectral analysis holds on the hypergroup K*.

Let K be a commutative hypergroup. The function $\varphi : K \to \mathbb{C}$ is called an *exponential*, if it is non-identically zero and satisfies

$$\varphi(x * y) = \varphi(x)\varphi(y)$$

for all x, y in K. Obviously, for any exponential φ the variety $\tau(\varphi)$ is of one dimension. Conversely, any one-dimensional variety is generated by an exponential. Hence spectral analysis holds for a given variety V if and only if it contains an exponential. According to Theorem 6.1 spectral analysis holds for a given variety if and only if its annihilator ideal is contained in a maximal ideal.

Let V be a variety in $C(K)$. We say that *spectral synthesis holds for* V, if V is the sum of finite dimensional varieties. This means that there is a set $(V_\gamma)_{\gamma \in \Gamma}$ of finite dimensional subvarieties of V such that any element f of V can be represented in the form

$$f = f_{\gamma_1} + f_{\gamma_2} + \cdots + f_{\gamma_n}$$

with some positive integer n, with some elements $\gamma_1, \gamma_2, \ldots, \gamma_n$ in Γ and with some functions f_{γ_i} in V_{γ_i}. If K is a locally compact Abelian group, convolution is defined by $\delta_x * \delta_y = \delta_{xy}$, involution is defined by $x^\vee = -x$ and e is the zero element of the group, then finite dimensional varieties are spanned by exponential monomials (see e.g. [Szé82a]), hence spectral synthesis holds for a variety in the hypergroup-sense if and only if the linear subspace of the variety spanned by its exponential monomials is dense in the variety.

If spectral synthesis holds for any variety in $C(K)$, then we say that *spectral synthesis holds on the hypergroup* K.

We shall see that – similarly to the case of groups – the Fourier–Laplace transform on $\mathcal{M}_c(K)$ can be used successfully in the study of spectral synthesis. If K is a commutative hypergroup and μ is an element of $\mathcal{M}_c(K)$, then for any exponential φ on K we define

$$\widehat{\mu}(\varphi) = \int_K \varphi(x^\vee) \, d\mu(x) \, .$$

Then $\widehat{\mu}$ is a complex valued function defined on the set of all exponentials on K. The mapping $\mu \mapsto \widehat{\mu}$ is obviously linear. It also has the important property

$$(\mu * \nu)\widehat{} = \widehat{\mu}\, \widehat{\nu}$$

for all μ, ν in $\mathcal{M}_c(K)$, which is called the *convolution formula* (see [BH95]).

6.3 Spectral analysis on polynomial hypergroups in a single variable

In this section we show that spectral analysis holds for any polynomial hypergroup in a single variable.

Theorem 6.2. *Spectral analysis holds for any polynomial hypergroup.*

Proof. Let K be the hypergroup associated with the sequence of polynomials $(P_n)_{n \in \mathbb{N}}$ and let V be an arbitrary variety in $\mathcal{C}(K)$. We remark that $\mathcal{C}(K)$ is the set of all complex valued functions on $\mathbb{N}$, equipped with the pointwise linear operations and with the topology of pointwise convergence. Accordingly, $\mathcal{M}_c(K)$ is the set of all finitely supported complex measures on $\mathbb{N}$. By Theorem 6.1 the annihilator $V^\perp$ of V is a proper closed ideal in $\mathcal{M}_c(K)$. By the convolution formula the Fourier–Laplace transforms of the elements of $V^\perp$ form a proper ideal in the ring of the Fourier–Laplace transforms of all elements of $\mathcal{M}_c(K)$. By Theorem 2.2 the set of all exponentials of K can be identified by $\mathbb{C}$. For any μ in $\mathcal{M}_c(K)$ and for any λ in $\mathbb{C}$ we have

$$\widehat{\mu}(\lambda) = \int_{\mathbb{N}} P_n(\lambda) \, d\mu(n) \,.$$

As μ is finitely supported, hence $\widehat{\mu}$ is a complex polynomial on $\mathbb{C}$. We can see easily that any complex polynomial on $\mathbb{C}$ can be written in the form $\widehat{\mu}$ with some μ in $\mathcal{M}_c(K)$. Indeed, if p is a complex polynomial on $\mathbb{C}$ of degree n, then it can be written in the form $p = \sum_{k=0}^n c_k P_k$ with some complex constants c_k $(k = 0, 1, \ldots, n)$. Then we have

$$p = \left(\sum_{k=0}^n c_k \delta_k \right)^{\widehat{}} \,.$$

This means that the Fourier–Laplace transforms of the elements of $V^\perp$ form a proper ideal in the ring of all complex polynomials on $\mathbb{C}$. It is known that in this case there exists a complex λ_0 such that $\widehat{\mu}(\lambda_0) = 0$ for all μ in $V^\perp$, which means, by definition, that the exponential $n \mapsto P_n(\lambda_0)$ belongs to V and the theorem is proved. □

6.4   Exponential polynomials on polynomial hypergroups in a single variable

In order to study spectral synthesis on hypergroups it is necessary to introduce a reasonable concept for exponential monomials. As the product of exponentials is not necessarily an exponential, it turns out that the above mentioned concept is not the appropriate one. In the case of polynomial hypergroups, however, we introduce the following concept. Let K be the polynomial hypergroup associated with the sequence of polynomials $(P_n)_{n \in \mathbb{N}}$. We call the function $\varphi : \mathbb{N} \to \mathbb{C}$ an *exponential monomial*, if it has the form

$$\varphi(n) = \sum_{j=0}^{k} c_j P_n^{(j)}(\lambda)$$

for each n in $\mathbb{N}$, where k is a nonnegative integer and λ, c_j $(j = 0, 1, \ldots, k)$ are complex numbers. A sum of exponential monomials is called an *exponential polynomial*. Now we list some basic properties of exponential monomials and polynomials on polynomial hypergroups. In what follows K is a fixed polynomial hypergroup associated with the sequence of polynomials $(P_n)_{n \in \mathbb{N}}$.

Theorem 6.3. *Let k be a nonnegative integer and λ a complex number. Then the functions $n \mapsto P_n^{(j)}$ $(j = 0, 1, \ldots, k)$ are linearly independent.*

Proof. Suppose that

$$c_0 P_n(\lambda) + c_1 P_n'(\lambda) + \cdots + c_k P_n^{(k)}(\lambda) = 0$$

holds for all n in $\mathbb{N}$ with some complex numbers c_j $(j = 0, 1, \ldots, k)$. Substituting $n = 0$ we have $c_0 = 0$. Supposing that we have proved $c_0 = c_1 = \cdots = c_m = 0$ for some $0 \leq m < k$, then substituting $n = m + 1$ we have $c_{m+1} = 0$. Hence, by induction, we have the statement. $\square$

Theorem 6.4. *Let V be a variety over K, k a positive integer, m_i a nonnegative integer and $\lambda_i, c_{i,j}$ complex numbers for $i = 1, 2 \ldots, k$ and $j = 0, 1, \ldots, m_i$. Suppose that $\lambda_s \neq \lambda_t$ for $s \neq t$. Let*

$$\varphi(n) = \sum_{i=1}^{k} \sum_{j=0}^{m_i} c_{i,j} P_n^{(j)}(\lambda_i)$$

for all n in $\mathbb{N}$. If φ belongs to V, then the function $n \mapsto c_{i,m_i} P_n^{(j)}(\lambda_i)$ belongs to V for $i = 1, 2, \ldots, k$ and $j = 0, 1, \ldots, m_i$.

Proof. Obviously we may suppose that $c_{i,m_i} \neq 0$ for $i = 1, 2, \ldots, k$. We prove the statement by induction on $m_1 + \cdots + m_k$. If $m_1 + \cdots + m_k = 0$, then

$$\varphi(n) = c_{1,0} P_n(\lambda_1) + \cdots + c_{k,0} P_n(\lambda_k)$$

holds for all n in $\mathbb{N}$. We have to show that the function $n \mapsto P_n(\lambda_i)$ belongs to V for $i = 1, 2, \ldots, k$. We prove this statement by induction on k. As it is obvious for $k = 1$ suppose that it has been proved for some $k \geq 1$ and let

$$\varphi(n) = c_{1,0} P_n(\lambda_1) + \cdots + c_{k,0} P_n(\lambda_k) + c_{k+1,0} P_n(\lambda_{k+1})$$

for all n in $\mathbb{N}$.

Let for all n in $\mathbb{N}$

$$\psi(n) = \varphi(n * 1) - \lambda_{k+1} \varphi(n) \, .$$

Then ψ belongs to V. On the other hand

$$\psi(n) = \sum_{i=1}^{k} c_{i,0}(\lambda_i - \lambda_{k+1}) P_n(\lambda_i)$$

holds for all n in $\mathbb{N}$ hence our statement follows.

Now suppose that the theorem is proved for some $m_1 + \cdots + m_k \geq 1$ and let

$$\varphi(n) = \sum_{i=1}^{k-1} \sum_{j=0}^{m_i} c_{i,j} P_n^{(j)}(\lambda_i) + \sum_{j=0}^{m_k} c_{k,j} P_n^{(j)}(\lambda_k) + c_{k,m_k+1} P_n^{(m_k+1)}(\lambda_k)$$

for all n in $\mathbb{N}$. (For $k = 1$ the first – empty – sum is zero.) Again, we set for all n in $\mathbb{N}$

$$\psi(n) = \varphi(n * 1) - \lambda_k \varphi(n).$$

Then ψ belongs to V. On the other hand

$$\psi(n) = \sum_{i=1}^{k-1} \sum_{j=0}^{m_i} c_{i,j} \left[\lambda_i P_n^{(j)}(\lambda_i) + j P_n^{(j-1)}(\lambda_i) - \lambda_k P_n^{(j)}(\lambda_i) \right]$$

$$+ \sum_{j=0}^{m_k} j c_{k,j} P_n^{(j-1)}(\lambda_k) + c_{k,m_k+1}(m_k+1) P_n^{(m_k)}(\lambda_k)$$

$$= \sum_{i=1}^{k} \sum_{j=0}^{m_k} b_{i,j} P_n^{(j)}(\lambda_i)$$

holds for all n in $\mathbb{N}$, where

$$b_{i,m_i} = c_{i,m_i}(\lambda_i - \lambda_k)$$

for $i = 1, 2, \ldots, k-1$ and

$$b_{k,m_k} = c_{k,m_k}(m_k + 1),$$

hence our statement follows.   □

A special case of this theorem is the following.

Theorem 6.5. *Let V be a variety over K, k a positive integer and λ a complex number. If the function $n \mapsto P_n^{(k)}(\lambda)$ belongs to V, then so do the functions $n \mapsto P_n^{(j)}(\lambda)$ for $j = 0, 1, \ldots, k-1$.*

Another important special case of Theorem 6.4 reads as follows.

Theorem 6.6. *Let K be the polynomial hypergroup associated with the sequence of polynomials $(P_n)_{n \in \mathbb{N}}$. If k is a positive integer, $m_1, m_2, \ldots, m_k$ are nonnegative integers and $\lambda_1, \lambda_2, \ldots, \lambda_k$ are different complex numbers, then the functions $n \mapsto P_n^{(j)}(\lambda_i)$ are linearly independent for $i = 1, 2, \ldots, k$ and $j = 0, 1, \ldots, m_i$.*

Proof.   Take $V = \{0\}$ in Theorem 6.4.   □

6.5   Spectral synthesis on polynomial hypergroups in a single variable

In this section we show that spectral synthesis holds for any polynomial hypergroup in the sense that the linear hull of all exponential monomials is dense in any variety. Actually, we shall prove that any variety on a polynomial hypergroup is finite dimensional and it is generated by functions of the form $n \mapsto P_n^{(j)}(\lambda)$ with some finite set of j's and some finite set of λ's. We use the notation of the previous section.

Theorem 6.7. *Spectral synthesis holds for any polynomial hypergroup.*

Proof.   Let $(P_n)_{n \in \mathbb{N}}$ be the sequence of polynomials with which the polynomial hypergroup K is associated. First we show that the variety generated by the function $n \mapsto P_n^{(k)}(\lambda)$ is finite dimensional for any nonnegative integer k and for any complex number λ. Let $\psi(n) = P_n^{(k)}(\lambda)$ for any n, k in $\mathbb{N}$ and λ in $\mathbb{C}$. Then by the linearization formula we have

$$\psi(n * m) = \sum_{j=0}^{k} \binom{k}{j} P_n^{(j)}(\lambda) P_m^{(k-j)}(\lambda)$$

for all m, n in $\mathbb{N}$, which yields the statement.

Now we know that for any variety V in $\mathcal{C}(K)$ the Fourier–Laplace transforms of the elements of $V^{\perp}$ form a proper ideal in the ring of all complex polynomials on $\mathbb{C}$. We denote this ideal by J. It is known that in this case there exist complex numbers $\lambda_1, \lambda_2, \ldots, \lambda_k$ and nonnegative integers $m_1, m_2, \ldots, m_k$ such that a polynomial p belongs to J if and only if $p^{(j)}(\lambda_i) = 0$ holds for $i = 1, 2, \ldots, k$ and $j = 0, 1, \ldots, m_i$. This means that the measure μ in $\mathcal{M}_c(K)$ annihilates V if and only if $\widehat{\mu}^{(j)}(\lambda_i) = 0$ holds for $i = 1, 2, \ldots, k$ and $j = 0, 1, \ldots, m_i$, that is, if and only if the functions $n \mapsto P_n^{(j)}(\lambda_i)$ are annihilated by μ for $i = 1, 2, \ldots, k$ and $j = 0, 1, \ldots, m_i$. It follows that V is the closure of the linear hull of these functions. As these functions generate finite dimensional varieties, our statement is proved. $\square$

6.6   Spectral analysis and spectral synthesis on multivariate polynomial hypergroups

In this section we will follow the ideas used above for the case of polynomial hypergroups in one variable to prove that spectral analysis and spectral synthesis holds for any multivariate polynomial hypergroup. The basic tools are Hilbert's Nullstellensatz (see e.g. [ZS75b]), the Noether–Lasker Theorem (see [ZS75a]) and the characterization of polynomial ideals in $\mathbb{C}[z_1, z_2, \ldots, z_n]$ by differential operators as it is presented in the Ehrenpreis–Palamodov Theorem and its consequences (see e.g. [Bjö79], [Stu02]).

First we define exponential monomials and exponential polynomials on multivariate polynomial hypergroups. Using the ideas in the single variable case the following concept seems to be the reasonable one: let K be a multivariate hypergroup in d dimensions associated with the family of polynomials $(Q_x)_{x \in K}$. For any polynomial P in d variables and for any ξ in $\mathbb{C}^d$ the function $x \mapsto P(\partial)Q_x(\xi)$ is called an *exponential monomial* on K and linear combinations of exponential monomials are called *exponential polynomials* on K. It is clear that in the case $d = 1$ we have the concept of exponential polynomials on polynomial hypergroups in a single variable as it was defined in the previous sections (see also in [Szé04]). By differentiating (3.1) one gets easily that the variety generated by an exponential polynomial is of finite dimension. If V is any variety in $\mathcal{C}(K)$ which contains an exponential, then we say that *spectral analysis holds for V*. If spectral analysis holds for any variety in $\mathcal{C}(K)$, then we say that *spectral analysis holds in K*. If V is any variety in $\mathcal{C}(K)$ and the linear hull of all exponential monomials in V is dense in V, then we say that *spectral synthesis holds for V*. If spectral synthesis holds for any variety in $\mathcal{C}(K)$, then we say that *spectral synthesis holds in K*.

For any finitely supported measure μ in $\mathcal{M}_c(K)$ the *Fourier–Laplace transform* $\widehat{\mu} : \mathbb{C}^d \to \mathbb{C}$ of μ is defined by the equation

$$\widehat{\mu}(\lambda) = \int_K Q_x(\lambda) \, d\mu(x)$$

for any λ in $\mathbb{C}^d$. Then $\widehat{\mu}$ is a polynomial in $\mathbb{C}[z_1, z_2, \ldots, z_n]$. On the other hand, the property of the generating family of polynomials $(Q_x)_{x \in K}$ guarantees that any polynomial in $\mathbb{C}[z_1, z_2, \ldots, z_n]$ is the Fourier–Laplace transform of some finitely supported measure μ in $\mathcal{M}_c(K)$. Indeed, if δ_x denotes the point mass concentrated at the element x in K and P is any polynomial in $\mathbb{C}[z_1, z_2, \ldots, z_n]$, then P can be represented in the form

$$P(z) = \sum_{i=1}^{n} c_i \, Q_{x_i}(z)$$

for any z in $\mathbb{C}^d$ with some $x_1, \ldots, x_n$ in K and complex numbers $c_1, c_2, \ldots, c_n$. As obviously $Q_{x_i} = \hat{\delta}_{x_i}$ for $i = 1, 2, \ldots, n$, hence

$$P = \sum_{i=1}^{n} c_i \, \hat{\delta}_{x_i} = \left(\sum_{i=1}^{n} c_i \, \delta_{x_i} \right)^{\widehat{}}.$$

If V is a variety in $\mathcal{C}(K)$, then its annihilator is a proper ideal in $\mathcal{M}_c(K)$ and the Fourier–Laplace transforms of the measures in this annihilator form a proper ideal in the polynomial ring $\mathbb{C}[z_1, z_2, \ldots, z_n]$. This follows from the well-known properties of convolution over hypergroups (see [BH95]). By Hilbert's Nullstellensatz there is a common root ξ of all these Fourier–Laplace transforms, that is,

$$\widehat{\mu}(\xi) = \int_K Q_x(\xi) \, d\mu(x) = 0$$

holds for each μ in the annihilator of V with some ξ in $\mathbb{C}^d$, which means that the exponential $x \mapsto Q_x(\xi)$ annihilates the annihilator of V, hence this exponential belongs to V. This proves the following theorem.

Theorem 6.8. *Spectral analysis holds for any multivariate polynomial hypergroup.*

In order to prove spectral synthesis for polynomial hypergroups in several variables first we need the following corollary of the Ehrenpreis–Palamodov Theorem (see Theorem 10.13 in [Stu02], p. 142.) We recall that for any ideal I in $\mathbb{C}[z_1, z_2, \ldots, z_n]$ the corresponding variety is the set of all common zeros of all polynomials in I.

Theorem 6.9. *Let I be a primary ideal in $\mathbb{C}[z_1, z_2, \ldots, z_n]$ and let V be the algebraic variety corresponding to I. Then there exist differential operators with polynomial coefficients*

$$A_i(z, \partial) = \sum_j c_{i,j} \cdot p_j(z_1, z_2, \ldots, z_n) \cdot \partial_1^{j_1} \partial_2^{j_2} \ldots \partial_n^{j_n}, \qquad i = 1, 2, \ldots, t$$

with the following property: a polynomial f in $\mathbb{C}[z_1, z_2, \ldots, z_n]$ lies in the ideal I if and only if the result applying $A_i(z, \partial)$ to f vanishes on V for $i = 1, 2, \ldots, t$.

We note the *algebraic variety* corresponding to an ideal in $\mathbb{C}[z_1, z_2, \ldots, z_n]$ is the set of all common zeros of all polynomials belonging to I. From this theorem we obtain immediately the following result.

Theorem 6.10. *Spectral synthesis holds for any multivariate polynomial hypergroup.*

Proof. Let K be a multivariate polynomial hypergroup and let V be any proper variety in $\mathcal{C}(K)$. Then the annihilator J of V is a proper closed ideal in $\mathcal{M}_c(K)$ and so is the set I of the Fourier–Laplace transforms of the elements of J in $\mathbb{C}[z_1, z_2, \ldots, z_n]$. By the Noether–Lasker Theorem I is the intersection of (finitely many) primary ideals, which means, by the above theorem, that for any ξ in V there is a set of polynomials $\mathcal{P}_\xi$ such that the finitely supported measure μ in $\mathcal{M}_c(K)$ annihilates V if and only if

$$P(\partial)\widehat{\mu}(\xi) = \int_K P(\partial)Q_x(\xi)\,d\mu(x) = 0$$

for each ξ in V and P in $\mathcal{P}_\xi$. This means that the exponential monomials $x \mapsto P(\partial)Q_x(\xi)$ for ξ in V and P in $\mathcal{P}_\xi$ belong to V and their linear hull is dense in V. $\qquad\square$

In particular, any finite dimensional translation invariant function space over a multivariate polynomial hypergroup consists of exponential polynomials. This result is closely related to some results in [Eng70], [McK76], [McK77b], [McK77a], [Szé82a] (see also [Szé91]).

6.7   Spectral analysis and moment functions

It turns out that moment functions play a special role in spectral analysis of varieties. Namely, if a variety contains nonzero moment functions, then spectral analysis holds on the variety. To show this, first we need some lemmas.

Lemma 5. *Let K be a commutative hypergroup, k a nonnegative integer, and let $(\varphi_j)_{j=0}^{k+1}$ be a generalized moment function sequence of nonzero functions. Then these functions are linearly independent. In particular, none of them in this sequence is included in the linear hull of the previous ones.*

Proof. Obviously φ_0 is nonzero. Suppose that $\varphi_1 = \lambda\varphi_0$ with some nonzero complex number λ. Then, by equation (1.19), we have

$$\varphi_0(x)\varphi_0(y) = \varphi_0(x * y) = \frac{1}{\lambda}\varphi_1(x * y)$$

$$= \frac{1}{\lambda}\varphi_1(x) \cdot \varphi_0(y) + \frac{1}{\lambda}\varphi_1(y) \cdot \varphi_0(x) = 2\varphi_0(x)\varphi_0(y),$$

which implies $\varphi_0 = 0$, a contradiction.

We may suppose that for $1 \leq j \leq k$ the function φ_j is not included in the linear hull of the previous ones and we assume that there exist complex numbers λ_j for $j = 0, 1, \ldots, k$ such that we have

$$\varphi_{k+1} = \sum_{j=0}^{k} \lambda_j\varphi_j. \tag{6.1}$$

By equation (1.19), we have

$$\varphi_{k+1}(x * y) = \sum_{j=0}^{k+1} \binom{k+1}{j}\varphi_j(x)\varphi_{k+1-j}(y) = \sum_{j=0}^{k} \lambda_j \sum_{i=0}^{j} \binom{j}{i}\varphi_i(x)\varphi_{j-i}(y) \tag{6.2}$$

for all x, y in K. Rearranging the terms we get

$$\sum_{j=0}^{k} \binom{k+1}{j}\varphi_j(x)\varphi_{k+1-j}(y) + \varphi_{k+1}(x)\varphi_0(y)$$

$$= \sum_{j=0}^{k}\sum_{i=j}^{k} \lambda_i \binom{i}{j}\varphi_j(x)\varphi_{i-j}(y).$$

Substituting φ_{k+1} from (6.1) it follows

$$\sum_{j=0}^{k}\left[\lambda_j\varphi_0(y) + \binom{k+1}{j}\varphi_{k+1-j}(y) - \sum_{i=j}^{k}\lambda_i \binom{i}{j}\varphi_{i-j}(y)\right]\varphi_j(x) = 0 \tag{6.3}$$

for each x, y in K. As $\varphi_k \neq 0$, hence, by our assumption, the coefficient of $\varphi_k(x)$ in equation (6.3) must be zero for each y in K, or else φ_k would be contained in the linear span of the functions $\varphi_0, \varphi_1, \ldots, \varphi_{k-1}$. That is, we have

$$0 = \lambda_k \varphi_0(y) + \binom{k+1}{k} \varphi_1(y) - \lambda_k \varphi_0(y) = (k+1)\varphi_1(y)$$

for each y in K, hence $\varphi_1 = 0$ and this is a contradiction. □

Theorem 6.11. *Let K be a commutative hypergroup, V a variety in (K), k a nonnegative integer and let $(\varphi_j)_{j=0}^{k+1}$ be a generalized moment function sequence of nonzero functions. If φ_{k+1} belongs to V, then φ_j belongs to V for $j = 0, 1, \ldots, k$.*

Proof. By the previous lemma the functions φ_j for $j = 0, 1, \ldots, k+1$ are linearly independent. Hence there exist elements x_i for $i = 0, 1, \ldots, k+1$ in K such that the matrix $\left(\varphi_j(x_i)\right)_{i,j=0}^{k+1}$ is regular. By equation (1.19) we have

$$\varphi_{k+1}(x * x_i) = \sum_{j=0}^{k+1} \binom{k+1}{j} \varphi_j(x)\varphi_{k+1-j}(x_i) \tag{6.4}$$

for all x in K and $i = 0, 1, \ldots, k+1$. For each x in K this is a system of linear equations for the unknowns $\binom{k+1}{j}\varphi_j(x)$, $j = 0, 1, \ldots, k+1$, the matrix of which is regular. Hence, by Cramer's Rule, φ_j, $j = 0, 1, \ldots, k+1$ is a linear combination of the translates $\tau_{x_i}\varphi_{k+1}$, $i = 0, 1, \ldots, k+1$, which belong to V. Our theorem is proved. □

Chapter 7

Spectral analysis and synthesis on Sturm–Liouville hypergroups

7.1 Exponential monomials on Sturm–Liouville hypergroups

The study of spectral analysis and spectral synthesis problems is based on the concept of exponential monomials. Unfortunately at this moment we do not have a general definition of this concept on arbitrary (commutative) hypergroups hence on each special type of hypergroups we need to introduce the most appropriate form. In Sections 6.4 and 6.6 we have seen the corresponding definition on polynomial hypergroups in one variable and in several variables, respectively. Using these concepts we were able to prove spectral synthesis on these types of hypergroups. Here we define exponential monomials on Sturm–Liouville hypergroups and we prove spectral analysis for finite dimensional varieties.

Let $K = (\mathbb{R}_0, A)$ be a Sturm–Liouville hypergroup. In Chapter 5 we described the exponential functions and an exponential family of Sturm–Liouville hypergroups. We recall that the continuous function $m : \mathbb{R}_0 \to \mathbb{C}$ is an exponential on K if and only if it is C^∞ on the positive reals and there exists a complex number λ such that

$$m''(x) + \frac{A'(x)}{A(x)} m'(x) = \lambda m(x), \qquad m(0) = 1, \qquad m'(0) = 0 \qquad (7.1)$$

holds for any positive x. We have seen that we can define an exponential family $\varphi : \mathbb{R}_0 \times \mathbb{C} \to \mathbb{C}$ with the property that the function $x \mapsto \varphi(x, \lambda)$ is an exponential of K for each complex λ and for each exponential m of K there exists a unique complex λ such that $m(x) = \varphi(x, \lambda)$ holds for every x in $\mathbb{R}_0$. Using this exponential family we define *exponential monomials* on K as functions of the form $x \mapsto P(\partial_2)\varphi(x,\lambda)$, where P is a complex polynomial and λ is a complex number. Here ∂_2 denotes the partial differential operator with respect to the second variable and the meaning of $P(\partial_2)$ is obvious. In particular, with $P \equiv 1$ we have that every exponential function is an exponential monomial. Observe that this is an analoguous concept to that we called an "exponential monomial" on polynomial hypergroups in several variables in Section 6.6. Sums of exponential monomials are called *exponential polynomials*.

A particular subclass of exponential monomials is formed by the functions of the type $x \mapsto \partial_2^k \varphi(x,\lambda)$, where k is a nonnegative integer and λ is a complex number. Here we note that if $\lambda = 0$, then $\varphi(x,0) = 1$ for each x in $\mathbb{R}_0$, hence the corresponding function $x \mapsto \partial_2^k \varphi(x,0)$ is identically 1 for $k = 0$, and it is identically 0 for $k > 0$. For the sake of simplicity we will call the functions $x \mapsto \partial_2^k \varphi(x,\lambda)$ *special exponential monomials* if k is a nonnegative integer and λ is a complex number, supposing that if $\lambda = 0$, then $k = 0$. Our aim is to show that different special exponential monomials are linearly independent and any finite dimensional variety in $\mathcal{C}(K)$ contains exponentials. In particular, spectral analysis holds for finite dimensional varieties.

7.2   Linear independence of special exponential monomials

First we show that different exponential functions are linearly independent.

Theorem 7.1. *On any Sturm–Liouville hypergroup different exponentials are linearly independent.*

Proof. We actually prove that different exponentials are linearly independent on *any* hypergroup.

Let $m_1, m_2, \ldots, m_n$ be different exponentials on the hypergroup K. We prove by induction on n. For $n = 1$ the statement is trivial. Suppose that $n > 1$ and

$$c_1 m_1(t) + c_2 m_2(t) + \cdots + c_{n-1} m_{n-1}(t) + c_n m_n(t) = 0 \qquad (7.2)$$

holds for each t in K. Let x, y be arbitrary in K and we integrate both sides of equation (7.2) with respect to the measure $\delta_x * \delta_y$:

$$c_1 m_1(x*y) + c_2 m_2(x*y) + \cdots + c_{n-1} m_{n-1}(x*y) + c_n m_n(x*y) = 0. \quad (7.3)$$

Using the exponential property of the m's we have

$$c_1 m_1(x) m_1(y) + \cdots + c_{n-1} m_{n-1}(x) m_{n-1}(y) + c_n m_n(x) m_n(y) = 0. \quad (7.4)$$

Now we write $t = x$ in (7.2) and multiply the equation obtained by $m_n(y)$:

$$c_1 m_1(x) m_n(y) + \cdots + c_{n-1} m_{n-1}(x) m_n(y) + c_n m_n(x) m_n(y) = 0. \quad (7.5)$$

We subtract (7.5) from (7.4) to get

$$c_1 m_1(x)[m_1(y) - m_n(y)] + \cdots + c_{n-1} m_{n-1}(x)[m_{n-1}(y) - m_n(y)] = 0. \quad (7.6)$$

By assumption the exponentials $m_1, m_2, \ldots, m_{n-1}$ are linearly independent, hence

$$c_i[m_i(y) - m_n(y)] = 0 \qquad (7.7)$$

for $i = 1, 2, \ldots, n - 1$. As $m_n \neq m_1$ we can choose a y in K such that $m_n(y) \neq m_1(y)$; it follows that $c_1 = 0$. Continuing this argument we get $c_i = 0$ for $i = 1, 2, \ldots, n - 1$, which also implies $c_n = 0$. The proof is complete. $\qquad \square$

We shall also need the following result in the sequel.

Theorem 7.2. *Let K be a Sturm–Liouville hypergroup with the exponential family $\varphi : \mathbb{R}_+ \times \mathbb{C} \to \mathbb{C}$, n a nonnegative integer and $\lambda_0 \neq 0$ a complex number. Then the special exponential monomials*

$$x \mapsto \varphi(x, \lambda_0), x \mapsto \partial_2 \varphi(x, \lambda_0), \ldots, x \mapsto \partial_2^n \varphi(x, \lambda_0) \qquad (7.8)$$

are linearly independent.

Proof.   We prove the statement by induction on n, which is obviously true for $n = 0$. Suppose that we have proved it for n and we prove it for $n + 1$, where n is some nonnegative integer. Proving our statement by contradiction we suppose that the function $x \mapsto \partial_2^{n+1}\varphi(x, \lambda_0)$ is a linear combination of the functions

$$x \mapsto \partial_2^k \varphi(x, \lambda_0)$$

for $k = 0, 1, \ldots, n$, that is, there are complex numbers c_k for $k = 0, 1, \ldots, n$ such that

$$\partial_2^{n+1}\varphi(x, \lambda_0) = \sum_{k=0}^{n} c_k \partial_2^k \varphi(x, \lambda_0) \tag{7.9}$$

holds for each x in K. In order to simplify our notation we let

$$p(x) = \frac{A'(x)}{A(x)}$$

for each $x > 0$. By the definition of the exponential family we have

$$\partial_1^2 \varphi(x, \lambda) + p(x)\partial_1 \varphi(x, \lambda) = \lambda \varphi(x, \lambda) \tag{7.10}$$

for each $x > 0$ and λ in $\mathbb{C}$. We differentiate both sides k times with respect to λ for $k = 0, 1, \ldots, n + 1$. We then obtain

$$\partial_1^2 \partial_2^k \varphi(x, \lambda) + p(x)\partial_1 \partial_2^k \varphi(x, \lambda) = \sum_{j=0}^{k} \binom{k}{j} \lambda^{(j)} \cdot \partial_2^{k-j} \varphi(x, \lambda), \tag{7.11}$$

or, equivalently,

$$\partial_1^2 \partial_2^k \varphi(x, \lambda) + p(x)\partial_1 \partial_2^k \varphi(x, \lambda) = \lambda \partial_2^k \varphi(x, \lambda) + k\partial_2^{k-1} \varphi(x, \lambda) \tag{7.12}$$

for each $x > 0$ and λ in $\mathbb{C}$ and for $k = 0, 1, \ldots, n+1$. (Here $\partial_2^{-1}\varphi(x, \lambda) = 0$.) We shall use this equation several times in the sequel.

Differentiating equation (7.9) two times with respect to x we have the equations

$$\partial_1 \partial_2^{n+1} \varphi(x, \lambda_0) = \sum_{k=0}^{n} c_k \partial_1 \partial_2^k \varphi(x, \lambda_0) \tag{7.13}$$

and

$$\partial_1^2 \partial_2^{n+1} \varphi(x, \lambda_0) = \sum_{k=0}^{n} c_k \partial_1^2 \partial_2^k \varphi(x, \lambda_0) \tag{7.14}$$

for each $x > 0$. From these equations by (7.12) we have

$$\sum_{k=0}^{n} c_k \partial_1^2 \partial_2^k \varphi(x, \lambda_0) + \sum_{k=0}^{n} c_k p(x) \partial_1 \partial_2^k \varphi(x, \lambda_0)$$

$$= \lambda_0 \partial_2^{n+1} \varphi(x, \lambda_0) + (n+1) \partial_2^n \varphi(x, \lambda_0)$$

$$= \sum_{k=0}^{n} \lambda_0 c_k \partial_2^k \varphi(x, \lambda_0) + (n+1) \partial_2^n \varphi(x, \lambda_0) \, .$$

We can reorder the terms in this equation to obtain

$$\sum_{k=0}^{n} c_k [\partial_1^2 \partial_2^k \varphi(x, \lambda_0) + p(x) \partial_1 \partial_2^k \varphi(x, \lambda_0) - \lambda_0 \partial_2^k \varphi(x, \lambda_0)]$$

$$= (n+1) \partial_2^n \varphi(x, \lambda_0) \, ,$$

or, equivalently, using again (7.12)

$$\sum_{k=1}^{n} k c_k \partial_2^{k-1} \varphi(x, \lambda_0) - (n+1) \partial_2^n \varphi(x, \lambda_0) = 0 \, . \tag{7.15}$$

But this is a contradiction, because equation (7.15) presents a nontrivial linear combination of linearly independent functions, which has the value zero. Hence the proof is complete. $\qquad \square$

Now we are in the position to prove linear independence of the special exponential monomials.

Theorem 7.3. *On any Sturm–Liouville hypergroup different special exponential monomials are linearly independent.*

Proof. We have to show that any finite set of special exponential monomials is linearly independent. First we suppose that this set does not include the special exponential monomial 1. We may suppose that this set consists of special exponential monomials of the form

$$x \mapsto \partial_2^l \varphi(x, \lambda_j)$$

for $l = 0, 1, \ldots, n$ and $j = 1, 2, \ldots, k$ with some restrictions on the nonnegative integer n and positive integer k. Actually, we shall consider two cases: in the first case we suppose that we have proved the linear independence of the functions

$$x \mapsto \partial_2^l \varphi(x, \lambda_j)$$

for $l = 0, 1, \ldots, n$ and $j = 1, 2, \ldots, k$, where n is a nonnegative integer and k is a positive integer and we show that the function $x \mapsto \partial_2^{n+1} \varphi(x, \lambda_1)$ is not a linear combination of them, while in the second case we suppose that we have proved the linear independence of the functions

$$x \mapsto \partial_2^l \varphi(x, \lambda_s), x \mapsto \partial_2^{n+1} \varphi(x, \lambda_t)$$

for $l = 0, 1, \ldots, n$, $s = 1, 2, \ldots, k$ and $t = 1, 2, \ldots, j$, where n is a nonnegative integer, $k \geq 2$ is a positive integer and j is a positive integer with $j \leq k - 1$ and we show that the function $x \mapsto \partial_2^{n+1} \varphi(x, \lambda_{j+1})$ is not a linear combination of them. It is easy to see that any other case can be reduced to these two cases (eventually, by renumbering the λ's). We apply induction again: in the first case the statement is clearly true for $n = 0$ and $k = 1$. Also, if $n = 0$ and k is arbitrary, then the statement follows from Theorem 7.1 and if $k = 1$ and n is arbitrary, then the statement follows from Theorem 7.8. Hence we can consider the first case and prove by contradiction: suppose that the function $x \mapsto \partial_2^{n+1} \varphi(x, \lambda_1)$ is a linear combination of the functions

$$x \mapsto \partial_2^l \varphi(x, \lambda_j)$$

for $l = 0, 1, \ldots, n$ and $j = 1, 2, \ldots, k$, where n is a nonnegative integer and k is a positive integer. This means that there are complex numbers $c_{l,j}$ for $l = 0, 1, \ldots, n$ and $j = 1, 2, \ldots, k$ such that

$$\partial_2^{n+1} \varphi(x, \lambda_1) = \sum_{l=0}^{n} \sum_{j=1}^{k} c_{l,j} \partial_2^l \varphi(x, \lambda_j) \tag{7.16}$$

holds for each $x > 0$. Differentiating two times with respect to x we get the equations

$$\partial_1 \partial_2^{n+1} \varphi(x, \lambda_1) = \sum_{l=0}^{n} \sum_{j=1}^{k} c_{l,j} \partial_1 \partial_2^l \varphi(x, \lambda_j) \tag{7.17}$$

and

$$\partial_1^2 \partial_2^{n+1} \varphi(x, \lambda_1) = \sum_{l=0}^{n} \sum_{j=1}^{k} c_{l,j} \partial_1^2 \partial_2^l \varphi(x, \lambda_j) \tag{7.18}$$

for each $x > 0$. From these equations, by (7.12), we have

$$\sum_{l=0}^{n} \sum_{j=1}^{k} c_{l,j} \partial_1^2 \partial_2^l \varphi(x, \lambda_j) + \sum_{l=0}^{n} \sum_{j=1}^{k} c_{l,j} p(x) \partial_1 \partial_2^l \varphi(x, \lambda_j)$$

$$= \lambda_1 \partial_2^{n+1} \varphi(x, \lambda_1) + (n+1) \partial_2^n \varphi(x, \lambda_1)$$

$$= \sum_{l=0}^{n} \sum_{j=1}^{k} \lambda_1 c_{l,j} \partial_2^l \varphi(x, \lambda_j) + (n+1) \partial_2^n \varphi(x \lambda_1).$$

We can reorder the terms in this equation to obtain

$$\sum_{l=0}^{n} \sum_{j=1}^{k} c_{l,j} [\partial_1^2 \partial_2^l \varphi(x, \lambda_j) + p(x) \partial_1 \partial_2^l \varphi(x, \lambda_j) - \lambda_1 \partial_2^l \varphi(x, \lambda_j)]$$

$$= (n+1)\partial_2^n \varphi(x, \lambda_1) \,,$$

or, equivalently, using again (7.12)

$$\sum_{l=1}^{n}\sum_{j=1}^{k} l c_{l,j} \partial_2^{l-1} \varphi(x, \lambda_j) - (n+1)\partial_2^n \varphi(x, \lambda_1) = 0 \,. \qquad (7.19)$$

But this is a contradiction, because equation (7.19) presents a nontrivial linear combination of linearly independent functions, which has the value zero. Hence the proof of our statement in the first case is complete.

Now we consider the second case and we prove again by contradiction: we suppose that we have proved the linear independence of the functions

$$x \mapsto \partial_2^l \varphi(x, \lambda_s), x \mapsto \partial_2^{n+1} \varphi(x, \lambda_t)$$

for $l = 0, 1, \ldots, n$, $s = 1, 2, \ldots, k$ and $t = 1, 2, \ldots, j$, where n is a nonnegative integer, $k \geq 2$ is a positive integer and j is a positive integer with $j \leq k - 1$ and we show that the function $x \mapsto \partial_2^{n+1} \varphi(x, \lambda_{j+1})$ is a linear combination of them. This means that there are complex numbers $c_{l,s}, d_t$ for $l = 0, 1, \ldots, n$ and $s = 1, 2, \ldots, k$, $t = 1, 2, \ldots, j$ such that

$$\partial_2^{n+1} \varphi(x, \lambda_{j+1}) = \sum_{l=0}^{n}\sum_{s=1}^{k} c_{l,s} \partial_2^l \varphi(x, \lambda_s) + \sum_{t=1}^{j} d_t \partial_2^{n+1} \varphi(x, \lambda_t) \qquad (7.20)$$

holds for each $x > 0$. Differentiating two times with respect to x we get the equations

$$\partial_1 \partial_2^{n+1} \varphi(x, \lambda_{j+1}) = \sum_{l=0}^{n}\sum_{s=1}^{k} c_{l,s} \partial_1 \partial_2^l \varphi(x, \lambda_s) + \sum_{t=1}^{j} d_t \partial_1 \partial_2^{n+1} \varphi(x, \lambda_t) \quad (7.21)$$

and

$$\partial_1^2 \partial_2^{n+1} \varphi(x, \lambda_{j+1}) = \sum_{l=0}^{n}\sum_{s=1}^{k} c_{l,s} \partial_1^2 \partial_2^l \varphi(x, \lambda_s) + \sum_{t=1}^{j} d_t \partial_1^2 \partial_2^{n+1} \varphi(x, \lambda_t) \quad (7.22)$$

for each $x > 0$. From these equations, by (7.12), we have

$$\sum_{l=0}^{n}\sum_{s=1}^{k} c_{l,s}\partial_1^2\partial_2^l\varphi(x,\lambda_s) + \sum_{t=1}^{j} d_t\partial_1^2\partial_2^{n+1}\varphi(x,\lambda_t)$$

$$+ \sum_{l=0}^{n}\sum_{s=1}^{k} c_{l,s}p(x)\partial_1\partial_2^l\varphi(x,\lambda_s) + \sum_{t=1}^{j} d_tp(x)\partial_1\partial_2^{n+1}\varphi(x,\lambda_t)$$

$$= \lambda_{j+1}\partial_2^{n+1}\varphi(x,\lambda_{j+1}) + (n+1)\partial_2^n\varphi(x,\lambda_{j+1})$$

$$= \sum_{l=0}^{n}\sum_{s=1}^{k} \lambda_{j+1}c_{l,s}\partial_2^l\varphi(x,\lambda_s) + \sum_{t=1}^{j} d_t\lambda_{j+1}\partial_2^{n+1}\varphi(x,\lambda_t) + (n+1)\partial_2^n\varphi(x,\lambda_{j+1})\,.$$

We can reorder the terms in this equation to obtain

$$\sum_{l=0}^{n}\sum_{s=1}^{k} c_{l,s}[\partial_1^2\partial_2^l\varphi(x,\lambda_s) + p(x)\partial_1\partial_2^l\varphi(x,\lambda_s) - \lambda_{j+1}\partial_2^l\varphi(x,\lambda_s)]$$

$$+ \sum_{t=1}^{j} d_t[\partial_1^2\partial_2^{n+1}\varphi(x,\lambda_t) + p(x)\partial_1\partial_2^{n+1}\varphi(x,\lambda_t) - \lambda_{j+1}\partial_2^{n+1}\varphi(x,\lambda_t)]$$

$$= (n+1)\partial_2^n\varphi(x,\lambda_{j+1})\,,$$

or, equivalently, using again (7.12)

$$\sum_{l=1}^{n}\sum_{s=1}^{k} lc_{l,s}\partial_2^{l-1}\varphi(x,\lambda_s) + \sum_{t=1}^{j} d_t(n+1)\partial_2^n\varphi(x,\lambda_t) \qquad (7.23)$$

$$+ \sum_{l=0}^{n}\sum_{s=1}^{k} c_{l,s}(\lambda_s - \lambda_{j+1})\partial_2^l\varphi(x,\lambda_s) + \sum_{t=1}^{j} d_t(\lambda_t - \lambda_{j+1})\partial_2^{n+1}\varphi(x,\lambda_t)$$

$$- (n+1)\partial_2^n\varphi(x,\lambda_{j+1}) = 0\,.$$

The term containing $\partial_2^n \varphi(x, \lambda_{j+1})$ does not appear in the first two sums, it appears with zero coefficient in the third sum, it does not appear in the fourth sum, hence its coefficient on the left hand side is $-(n+1) \neq 0$. This is a contradiction and the proof of our statement also in the second case is complete.

To finish the proof we have to consider the case where the special exponential monomial 1 is in the set of the exponential monomials. We prove by contradiction again: suppose that there are nonzero complex numbers $\lambda_1, \lambda_2, \ldots, \lambda_k$ and there is a nonnegative integer n such that

$$1 = \sum_{l=0}^{n} \sum_{j=1}^{k} c_{l,j} \partial_2^l \varphi(x, \lambda_j) \tag{7.24}$$

holds for each $x > 0$. Differentiating equation two times with respect to x we obtain

$$0 = \sum_{l=0}^{n} \sum_{j=1}^{k} c_{l,j} \partial_1 \partial_2^l \varphi(x, \lambda_j) \tag{7.25}$$

and

$$0 = \sum_{l=0}^{n} \sum_{j=1}^{k} c_{l,j} \partial_1^2 \partial_2^l \varphi(x, \lambda_j) \tag{7.26}$$

for each $x > 0$. Adding equations (7.25) and (7.26) we get

$$0 = \sum_{l=0}^{n} \sum_{j=1}^{k} c_{l,j} [\partial_1^2 \partial_2^l \varphi(x, \lambda_j) + p(x) \partial_1 \partial_2^l \varphi(x, \lambda_j)] \tag{7.27}$$

$$= \sum_{l=0}^{n} \sum_{j=1}^{k} c_{l,j} [\lambda_j \partial_2^l \varphi(x, \lambda_j) + l \partial_2^{l-1} \varphi(x \lambda_j)]$$

for each $x > 0$. On the right hand side we have a linear combination of linearly independent functions. The coefficient of $\partial_2^n \varphi(x, \lambda_j)$ is $c_{n,j} \lambda_j$, which must be zero, hence $c_{n,j} = 0$ for $j = 1, 2, \ldots, k$. Continuing recursively we get that $c_{n-1,j} = c_{n-2,j} = \cdots = c_{0,j} = 0$ for $j = 1, 2, \ldots, k$, whih contradicts to equation (7.24). Now the proof of the theorem is complete.

$\square$

Here we prove another linear independence property of moment functions.

Lemma 6. *Let N be a positive integer. On a Sturm–Liouville hypergroup a nonzero linear combination of nondegenerate moment functions of order N is not a linear combination of moment functions of lower order.*

Proof. We prove by contradiction. Suppose that a nonzero linear combination of moment functions of order N is a linear combination of moment functions of lower order. By Theorem 4.6 a nondegenerate moment function of order N is a linear combination of functions of the form $x \mapsto \partial_2^j \varphi(x, \lambda)$, where $j = 1, 2, \ldots, N$, λ is a complex number and φ is the exponential family of the Sturm–Liouville hypergroup. As the moment functions of order N are nondegenerate, our assumption implies that there are complex numbers a_i and different complex numbers λ_i for $i = 1, 2, \ldots, k$ such that the function

$$x \mapsto a_1 \partial_2^N \varphi(x, \lambda_1) + a_2 \partial_2^N \varphi(x, \lambda_2) + \cdots + a_k \partial_2^N \varphi(x, \lambda_k)$$

is a linear combination of moment functions of order at most $N - 1$, moreover not all the coefficients a_i are equal to zero. We may suppose that a_1 is nonzero. Then the special exponential monomial $x \mapsto \partial_2^N \varphi(x, \lambda_1)$ is a linear combination of the special exponential monomials $x \mapsto \partial_2^N \varphi(x, \lambda_2), \ldots, \partial_2^N \varphi(x, \lambda_k)$ and some special exponential monomials of the form $x \mapsto \partial_2^j \varphi(x, \mu_j)$ with $j = 0, 1, \ldots, N-1$ and some complex numbers μ_j. But this contradicts to the linear independence of different special exponential monomials proved in Theorem 7.3. Our theorem is proved. $\square$

7.3   Spectral analysis on Sturm–Liouville hypergroups

In this section we show that spectral analysis holds for finite dimensional varieties on Sturm–Liouville hypergroups.

Theorem 7.4. *Spectral analysis holds for nonzero finite dimensional varieties on every Sturm–Liouville hypergroup.*

Proof. Suppose that K is a Sturm–Liouville hypergroup and $V \neq \{0\}$ is a finite dimensional variety in $\mathcal{C}(\mathbb{R})$. We have to show that V contains an exponential. Let $f_1, f_2, \ldots, f_n$ be a basis of V, then there exist complex valued functions $c_{i,j}$ for $i, j = 1, 2, \ldots, n$ such that

$$f_i(x * y) = \sum_{j=1}^{n} c_{i,j}(y) f_j(x) \tag{7.28}$$

holds for every x, y in K and $i = 1, 2, \ldots, n$. As the functions $f_1, f_2, \ldots, f_n$ are linearly independent, hence there are elements x_k for $k = 1, 2, \ldots, n$ in K such that the matrix $\left(f_j(x_k)\right)_{j,k=1}^{n}$ is regular. We have

$$f_i(x_k * y) = \sum_{j=1}^{n} c_{i,j}(y) f_j(x_k)$$

for each y in K and $k = 1, 2, \ldots, n$. For any fixed i this is an inhomogeneous system of linear equations for the unknowns $c_{i,j}(y)$ $(j = 1, 2, \ldots, n)$ with regular fundamental matrix, hence, by Cramer's Rule, $c_{i,j}$ is a linear combination of translates of f_i, hence $c_{i,j}$ belongs to V $(i, j = 1, 2, \ldots, n)$.

Going back to equation (7.28) and using the associativity of convolution we infer that

$$\sum_{j=1}^{n} c_{i,j}(z) f_j(x * y) = \sum_{j=1}^{n} c_{i,j}(y * z) f_j(x), \tag{7.29}$$

or

$$\sum_{j=1}^{n} c_{i,j}(z) \sum_{l=1}^{n} c_{j,l}(y) f_l(x) = \sum_{j=1}^{n} c_{i,j}(y * z) f_j(x) \tag{7.30}$$

holds for each x, y, z in K. This is equivalent to

$$\sum_{k=1}^{n} \sum_{j=1}^{n} c_{i,j}(z) c_{j,k}(y) f_k(x) = \sum_{k=1}^{n} c_{i,k}(y * z) f_k(x). \tag{7.31}$$

By the linear independence of the f_k's we have

$$\sum_{j=1}^{n} c_{i,j}(z) c_{j,k}(y) = c_{i,k}(y * z) = c_{i,k}(z * y) \tag{7.32}$$

for each y, z in K. Let $C(x)$ be the matrix $\left(c_{i,j}(x)\right)_{i,j=1}^{n}$, then from (7.32) it follows

$$C(x * y) = C(x) \cdot C(y) \qquad (7.33)$$

for each x, y in K. In particular, the matrices $C(x)$ are commuting for different x's. It follows that there exists a nonsingular matrix S such that the matrix $T(x)$ defined by

$$T(x) = S^{-1} \cdot C(x) \cdot S \qquad (7.34)$$

is lower triangular for each x in K. On the other hand, the entries of $T(x)$ are linear combinations of the $c_{i,j}$'s, hence they belong to V. Further we have for each x, y in K:

$$T(x * y) = S^{-1} \cdot C(x * y) \cdot S = S^{-1} \cdot C(x) \cdot C(y) \cdot S$$

$$= S^{-1} \cdot C(x) \cdot S \cdot S^{-1} \cdot C(y) \cdot S = T(x) \cdot T(y) \,.$$

Suppose that $T(x) = \left(t_{i,j}(x)\right)_{i,j=1}^{n}$, then using the fact that $T(x)$ is lower triangular we have that

$$t_{i,j}(x * y) = \sum_{k=j}^{i} t_{i,k}(x) \cdot t_{k,j}(y)$$

holds for $j = 1, 2, \ldots, i$ and for each x, y in K. If we put $j = i$ we get

$$t_{i,i}(x * y) = t_{i,i}(x) \cdot t_{i,i}(y) \qquad (7.35)$$

for $i = 1, 2, \ldots, n$ and for each x, y in K, which means that the functions $t_{i,i}$ $(i = 1, 2, \ldots, n)$ are exponentials in V and the theorem is proved. $\square$

We note that it is an open problem if spectral synthesis holds for finite dimensional varieties on any Sturm–Liouville hypergroup.

Observe that in the proof of the above theorem we have never used the fact that K is a Sturm–Liouville hypergroup: actually only the associativity and commutativity have been used. Hence we proved the following result.

Theorem 7.5. *Spectral analysis holds for nonzero finite dimensional varieties on any commutative hypergroup.*

Chapter 8

Moment problems on hypergroups

8.1 The moment problem in general

The classical *moment problem* published in 1894 by Thomas Jan Stieltjes (see [Sti94]) is the following: Given a sequence $s_0, s_1, \ldots$ of real numbers. Find necessary and sufficient conditions for the existence of a measure μ on $[0, \infty[$ so that

$$s_n = \int_0^\infty x^n \, d\mu(x)$$

holds for $n = 0, 1, \ldots$. Another form of the moment problem, also called "*Hausdorff's moment problem*" or the "*little moment problem*", may be stated as follows: Given a sequence of numbers $(s_n)_{n=0}^\infty$, under what conditions is it possible to determine a function α of bounded variation in the interval $]0, 1[$ such that

$$s_n = \int_0^1 x^n \, d\alpha(x)$$

for $n = 0, 1, \ldots$. Such a sequence is called a *moment sequence* and Felix Hausdorff (see [Hau21a], [Hau21b]) was the first to obtain necessary and sufficient conditions for a sequence to be a moment sequence. In both cases the question of uniqueness of μ, respectively α arise. For a detailed discussion on classical moment problems see e.g. [Akh65].

Let μ be a compactly supported measure on K and let $(\varphi_k)_{k=0}^\infty$ be a generalized moment function sequence. Then for each natural number n the complex number

$$m_n = \int_K \varphi_n \, d\mu$$

is called the *n-th generalized moment of* μ with respect to the given generalized moment function sequence. In this case the sequence $(m_n)_{n=0}^{\infty}$ is called the *generalized moment sequence of the measure* μ with respect to the given generalized moment function sequence.

In this setting we can formulate the problem of existence: Let the generalized moment function sequence $(\varphi_k)_{k=0}^{\infty}$ and the sequence of complex numbers $(m_n)_{n=0}^{\infty}$ be given. Under what conditions is there a compactly supported measure μ on K such that $(m_n)_{n=0}^{\infty}$ is the generalized moment sequence of the measure μ with respect to the given generalized moment function sequence? The other basic question is about the uniqueness: if the compactly supported measures μ and ν have the same generalized moment sequences with respect to the given generalized moment function sequence, then does it follow $\mu = \nu$?

8.2   Uniqueness on polynomial hypergroups

In this section we shall use the results in Section 2.3 on the representation of generalized moment functions on polynomial hypergroups in the following form (see Theorem 2.5).

Let $K = (\mathbb{N}, P_n)$ be a polynomial hypergroup and let $(\varphi_k)_{k=0}^{\infty}$ be a generalized moment function sequence on K. Then there exists a sequence $(c_k)_{k=0}^{\infty}$ such that for each natural number N we have

$$\varphi_k(n) = \left(P_n \circ f\right)^{(k)}(0) \quad (k = 0, 1, \ldots, N), \tag{8.1}$$

where

$$f(t) = \sum_{j=0}^{N} c_j \frac{t^j}{j!}$$

for each t in $\mathbb{R}$.

Theorem 8.1. *Let Let $K = (\mathbb{N}, P_n)$ be a polynomial hypergroup, μ a finitely supported measure on $\mathbb{N}$ and let $(\varphi_k)_{k=0}^{\infty}$ be a generalized moment function sequence on K. If $\varphi_1 \neq 0$ and*

$$\int_{\mathbb{N}} \varphi_k(n) \, d\mu(n) = 0 \qquad (8.2)$$

for $k = 0, 1, 2, \ldots$, then $\mu = 0$.

Proof. First we remark that

$$\int_{\mathbb{N}} \varphi_k(n) \, d\mu(n) = \sum_{n=0}^{N} \varphi_k(n) \mu_n \,. \qquad (8.3)$$

Hence, by assumption, we have the following system of equations

$$\sum_{n=0}^{N} \varphi_k(n) \mu_n = 0 \qquad (8.4)$$

for $k = 0, 1, 2, \ldots, N$.

On the other hand, by the result quoted from [OS05], we have that

$$\varphi_k(n) = \left(P_n \circ f \right)^{(k)}(0) \qquad (8.5)$$

holds for $k = 0, 1, 2, \ldots, N$, $n = 0, 1, 2, \ldots, N$, where

$$f(t) = \sum_{i=0}^{N} c_i \frac{t^i}{i!}$$

is a polynomial. Let $\lambda = f(0)$. From (8.5) we have for $k = 1$

$$\varphi_1(n) = P_n'(\lambda) \, c_1 \,,$$

which implies $c_1 \neq 0$.

Let n be a fixed nonnegative integer and we let for $k = 0, 1, 2, \ldots, N$ and for each t in $\mathbb{R}$

$$F_k(t) = \left(P_n \circ f \right)^{(k)}(t) \,. \qquad (8.6)$$

We show that for $k = 0, 1, 2, \ldots, N$ and for each t in $\mathbb{R}$

$$F_k(t) = \sum_{j=0}^{k} p_{k,j}(t) P_n^{(j)}\big(f(t)\big), \tag{8.7}$$

where $p_{k,j}$ is a polynomial and $p_{k,k}(t) = f'(t)^k$.

We prove (8.7) by induction on k. For $k = 0$ the statement is trivial with $p_{0,0}(t) = 1$. Supposing that (8.7) is proved we show it for $k+1$ instead of k. We then have

$$F_{k+1}(t) = F_k'(t) = \sum_{j=0}^{k} p_{k,j}'(t) P_n^{(j)}\big(f(t)\big) + \sum_{j=0}^{k} p_{k,j}(t) P_n^{(j+1)}\big(f(t)\big) f'(t),$$

and this is of the form (8.7) with $k + 1$ for k. Moreover, $p_{k+1,k+1}(t) = p_{k,k}(t) \cdot f'(t) = f'(t)^{k+1}$. Then, using (8.5), we have

$$\varphi_k(n) = \sum_{j=0}^{k} c_{k,j} P_n^{(j)}(\lambda) \quad k = 0, 1, 2, \ldots, N, \tag{8.8}$$

where $c_{k,k} = f'(0)^k \neq 0$, $c_{0,0} = 1$. By (8.4) it follows

$$\sum_{n=0}^{N} \sum_{j=0}^{k} c_{k,j} P_n^{(j)}(\lambda)\mu_n = 0 \tag{8.9}$$

for $k = 0, 1, 2, \ldots, N$. For $k = 0$ this means

$$\sum_{n=0}^{N} P_n(\lambda)\mu_n = 0. \tag{8.10}$$

Now let $k = 1$ in (8.9), then we have, by (8.10),

$$\sum_{n=0}^{N} c_{1,0} P_n(\lambda)\mu_n + c_{1,1} P_n'(\lambda)\mu_n = c_{1,0} \sum_{n=0}^{N} P_n(\lambda)\mu_n + c_{1,1} \sum_{n=0}^{N} P_n'(\lambda)\mu_n$$

$$= c_{1,1} \sum_{n=0}^{N} P'_n(\lambda)\mu_n = 0 \,.$$

As $c_{1,1} \neq 0$, then it follows

$$\sum_{n=0}^{N} P'_n(\lambda)\mu_n = 0 \,.$$

Continuing this process we get the system of equations

$$\sum_{n=0}^{N} P_n^{(k)}(\lambda)\mu_n = 0 \tag{8.11}$$

for $k = 0, 1, 2, \ldots, N$. Observe that the degree of P_n is exactly n, hence we can rewrite (8.11) in the form

$$\sum_{n=k}^{N} P_n^{(k)}(\lambda)\mu_n = 0 \tag{8.12}$$

for $k = 0, 1, 2, \ldots, N$. This is a homogeneous system of linear equations for the unknowns μ_n, $n = 0, 1, \ldots, N$. The fundamental matrix of this system is an $N \times N$ upper triangular matrix with the nonzero numbers $P_k^{(k)}(\lambda)$ in the main diagonal, hence this matrix is regular, which means that the system has only trivial solution: $\mu_n = 0$ for $n = 0, 1, 2, \ldots, N$. This means $\mu = 0$ and the proof is complete. □

This result obviously implies the following uniqueness theorem.

Theorem 8.2. *Let Let $K = (\mathbb{N}, P_n)$ be a polynomial hypergroup, μ, ν finitely supported measures on $\mathbb{N}$ and let $(\varphi_k)_{k=0}^{\infty}$ be a generalized moment function sequence on K. If $\varphi_1 \neq 0$ and the generalized moment sequences of μ and ν with respect to the given generalized moment function sequence are the same, then $\mu = \nu$.*

8.3   The case of Sturm–Liouville hypergroups

Following the previous ideas in this section we consider the same problem on Sturm–Liouville hypergroups. We shall use the results in Section 4.3 on the representation of generalized moment functions on Sturm–Liouville hypergroups in the following form (see [OS08]).

Let $K = (\mathbb{R}_0, A)$ be a Sturm–Liouville hypergroup and let Φ the exponential family of the hypergroup K (see [OS08]). This means that for each z in $\mathbb{C}$ and x in $\mathbb{R}_+$ the function Φ satisfies

$$\partial_1^2 \Phi(x, z) + \frac{A'(x)}{A(x)}\, \partial_1 \Phi(x, z) = z\, \Phi(x, z)\,, \qquad (8.13)$$

further $\Phi(0, z) = 1$ and $\partial_1 \Phi(0, z) = 0$. It follows that the function $z \mapsto \Phi(x, z)$ is entire for each x in $\mathbb{R}_0$. Let $(\varphi_k)_{k=0}^{\infty}$ be a generalized moment function sequence on K. Then there exists a sequence $(c_k)_{k=0}^{\infty}$ such that for each natural number N we have

$$\varphi_k(x) = \frac{d^k}{dt^k} \Phi\big(x, f(t)\big)(0) \qquad (8.14)$$

for $k = 0, 1, 2, \ldots N$, x in $\mathbb{R}_0$, t in $\mathbb{C}$, where

$$f(t) = \sum_{i=0}^{N} c_i\, \frac{t^i}{i!}$$

is a polynomial.

Theorem 8.3. *Let $K = (\mathbb{R}_0, A)$ be a Sturm–Liouville hypergroup, μ a compactly supported measure on $\mathbb{R}_0$ and let $(\varphi_k)_{k=0}^{\infty}$ be a generalized moment function sequence on K. If $\varphi_1 \neq 0$ and*

$$\int_{\mathbb{R}_0} \varphi_k(x)\, d\mu(x) = 0 \qquad (8.15)$$

for $k = 0, 1, 2, \ldots$, then $\mu = 0$.

Proof. We show that if

$$\int_{\mathbb{R}_0} \varphi_k(x) \, d\mu(x) = 0 \qquad (8.16)$$

for $k = 0, 1, 2, \ldots$, then $\mu = 0$.

Let N be a fixed positive integer. By the result quoted above from [OS08] we have

$$\varphi_k(x) = \frac{d^k}{dt^k} \Phi\big(x, f(t)\big)(0) \qquad (8.17)$$

for $k = 0, 1, 2, \ldots, N$ x in $\mathbb{R}_0$, t in $\mathbb{C}$, where

$$f(t) = \sum_{i=0}^{N} c_i \frac{t^i}{i!}$$

is a polynomial. Let $\lambda = f(0)$. From (8.17) we have for $k = 1$

$$\varphi_1(x) = \frac{d}{dt} \Phi\big(x, f(t)\big)(0) = \partial_2 \Phi(x, \lambda) c_1 \,,$$

which implies $c_1 \neq 0$.

Let x be a fixed nonnegative real number and we let for $k = 0, 1, 2, \ldots, N$ and for each t in $\mathbb{R}$

$$F_k(t) = \frac{d^k}{dt^k} \Phi\big(x, f(t)\big) \,. \qquad (8.18)$$

We show that for $k = 0, 1, 2, \ldots, N$ and for each t in $\mathbb{R}$

$$F_k(t) = \sum_{j=0}^{k} p_{k,j}(t) \partial_2^{(j)} \Phi\big(x, f(t)\big) \,, \qquad (8.19)$$

where $p_{k,j}$ is a polynomial and $p_{k,k}(t) = f'(t)^k$.

We prove (8.19) by induction on k. For $k = 0$ the statement is tivial. Supposing that (8.19) is proved we show it for $k + 1$ instead of k. We then have

$$F_{k+1}(t) = F_k'(t) = \sum_{j=0}^{k} p_{k,j}'(t)\partial_2^{(j)}\Phi\big(x, f(t)\big) + \sum_{j=0}^{k} p_{k,j}(t)\partial_2^{(j+1)}\Phi\big(x, f(t)\big)\, f'(t)$$

and this is the form (8.19) with $k+1$ for k. Moreover, $p_{k+1,k+1}(t) = p_{k,k}(t) \cdot f'(t) = f'(t)^k$.

Then, using (8.17), we have

$$\varphi_k(x) = \sum_{j=0}^{k} c_{k,j}\partial_2^{(j)}\Phi\big(x, \lambda\big) \quad k = 0, 1, 2, \ldots, N, \qquad (8.20)$$

where $c_{k,k} \neq 0$, $c_{0,0} = 1$.

By (8.16) it follows

$$\sum_{j=0}^{k} c_{k,j} \int_{\mathbb{R}_0} \partial_2^{(j)}\Phi\big(x, \lambda\big)\, d\mu(x) = 0 \qquad (8.21)$$

for $k = 0, 1, 2, \ldots$. For $k = 0$ this gives

$$\int_{\mathbb{R}_0} \Phi(x, \lambda)\, d\mu(x) = 0\,. \qquad (8.22)$$

For $k = 1$ we have

$$c_{1,0} \int_{\mathbb{R}_0} \Phi(x, \lambda)\, d\mu(x) + c_{1,1} \int_{\mathbb{R}_0} \partial_2\Phi\big(x, \lambda\big)\, d\mu(x) = 0\,. \qquad (8.23)$$

By (8.22) and $c_{1,1} \neq 0$ this implies

$$\int_{\mathbb{R}_0} \partial_2\Phi(x, \lambda)\, d\mu(x) = 0\,. \qquad (8.24)$$

Continuing this process we arrive at

$$\int_{\mathbb{R}_0} \partial_2^{(k)}\Phi(x, \lambda)\, d\mu(x) = 0 \qquad (8.25)$$

for $k = 0, 1, 2, \ldots, N$. As N is arbitrary, we actually have that (8.25) holds for $k = 0, 1, \ldots$.

We recall that the function $\widehat{\mu} : \mathbb{C} \to \mathbb{C}$ defined for each complex z by

$$\widehat{\mu}(z) = \int_{\mathbb{R}_0} \Phi(x, z) \, d\mu(x) \qquad (8.26)$$

is the Fourier–Laplace transform of the measure μ (see [BH95]). As μ is compactly supported, $\widehat{\mu}$ is an entire function. On the other hand, as the integration in (8.26) is performed on the compact support of μ and $z \mapsto \Phi(x, z)$ is an entire function, hence the differentiation and the integration in (8.23) can be interchanged. This means that we have that

$$\widehat{\mu}^{(k)}(z) = \frac{d^k}{dz^k} \int_{\mathbb{R}_0} \Phi(x, z) \, d\mu(x) = \int_{\mathbb{R}_0} \partial_2^{(k)} \Phi(x, z) \, d\mu(x) \qquad (8.27)$$

holds for $k = 0, 1, 2, \ldots$, and for all z in $\mathbb{C}$. In particular, for $z = \lambda$

$$\widehat{\mu}^{(k)}(\lambda) = \int_{\mathbb{R}_0} \partial_2^{(k)} \Phi(x, z) \, d\mu(x)\Big|_{z=\lambda} = \int_{\mathbb{R}_0} \partial_2^{(k)} \Phi(x, \lambda) \, d\mu(x) = 0 . \quad (8.28)$$

As $\widehat{\mu}$ is an entire function, it follows $\widehat{\mu} = 0$. Then $\mu = 0$ and our statement is proved. $\qquad \square$

Similarly as above, we have the corresponding uniqueness result.

Theorem 8.4. *Let* $K = (\mathbb{R}_0, A)$ *be a Sturm–Liouville hypergroup,* μ, ν *compactly supported measures on* $\mathbb{R}_0$ *and let* $(\varphi_k)_{k=0}^{\infty}$ *be a generalized moment function sequence on* K. *If* $\varphi_1 \neq 0$ *and the generalized moment sequences of* μ *and* ν *with respect to the given generalized moment function sequence are the same, then* $\mu = \nu$.

8.4 An approximation result

Using the results of Sections 5.2 and 5.4 we can extend the previous results to two-point support hypergroups. In both cases studied in Sections 5.2 and 5.4 we had that in the case $\varphi_1 \neq 0$ the corresponding generalized moment function sequence can be obtained from the corresponding exponential family exactly in the same manner as in the previous section. Hence we can infer that the same conclusion holds: the uniqueness of the moment problem follows for compactly supported measures. We have the following theorem.

Theorem 8.5. *Let K be either of the two-point support hypergroups studied in Sections 5.2 and 5.4 and let μ, ν be compacly supported measures on K. If $(\varphi_n)_{n \in \mathbb{N}}$ is a generalized moment function sequence on K with $\varphi_1 \neq 0$ and*

$$\int_K \varphi_n \, d\mu = \int_K \varphi_n \, d\nu \,,$$

holds for $n = 0, 1, \ldots$, then $\mu = \nu$.

Proof. For the proof it is enough to remark that the condition $\varphi_1 \neq 0$ excludes the case $\varphi_0 = 1$ and, by Theorem 5.7, we can apply the same method as in Theorem 8.3. □

By the Hahn–Banach theorem, Theorems 8.1 and 8.3 imply the following Weierstrass–type result.

Theorem 8.6. *Let K be either a polynomial, or a Sturm–Liouville hypergroup, or any of the two hypergroups studied in Sections 5.2 and 5.4, further let $(\varphi_n)_{n=0}^{\infty}$ be a generalized moment function sequence on K. If $\varphi_1 \neq 0$, then the linear hull of the sequence $(\varphi_n)_{n=0}^{\infty}$ is dense in $\mathcal{C}(K)$.*

We note that the assumption on φ_1 excludes the case of the trivial generalized moment function sequence: $\varphi_n = 0$ for $n = 1, 2, \ldots$, in which case the theorem trivially fails to hold.

Chapter 9

Special functional equations on hypergroups

9.1 The sine functional equation on polynomial hypergroups

This chapter is devoted to some special functional equations on commutative hypergroups. In the first two sections we solve the sine and the cosine functional equations on arbitrary polynomial hypergroups in a single variable. These equations have a long history and they have been studied by several authors (see e.g. [Acz66], [CKN85], [RS60], [Seg63], [Szé91], [Cor77], [Ste96]) under various assumptions. The method of solution here is based on spectral synthesis (see Section 6.3). Actually, this is an illustration of the strength of spectral analytical methods which have been introduced in the theory of functional equations started with the volume [Szé91]. It seems that this method can be effectively used also on hypergroups in the presence of spectral synthesis. The results presented here are taken from [Oro06b].

The results of the third section in this capter are also closely related to spectral synthesis. Indeed, we consider finite dimensional varieties on a commutative hypergroup. This leads to the study of the *Lévi–Cività functional equation*, which plays a basic role in differential geometry and representation theory. This equation has been studied by several mathematicians, too (see e.g. [Acz66], [AY10], [Szé91], [Shu11], [Los91], [Szé88]). In the case of this equation we do not assume the presence of spectral synthesis but, instead, we suppose that some special function family generates a dense subspace in the varieties on the underlying hypergroup. This function family arises from an exponential family in a quite natural way. This assumption is definitely satisfied if spectral synthesis is valid on the hypergroup, for instance, if it is polynomial. Nevertheless, it is an open question if our assumption implies spectral synthesis in some sense in general.

Now we turn to the study of the sine equation on polynomial hypergroups.

Theorem 9.1. *Let* $(\mathbb{N}, *)$ *be the polynomial hypergroup associated with the sequence of polynomials* $(P_n)_{n \in \mathbb{N}}$ *and let* $f, g : \mathbb{N} \to \mathbb{C}$ *be functions such that* f *is not identically zero. Then* f *and* g *satisfy the sine functional equation, that is,*

$$f(n * m) = f(n)g(m) + f(m)g(n) \tag{9.1}$$

holds for all m,n in $\mathbb{N}$ *if and only if* f *and* g *can be written in one of the following forms:*

Case I.

$$f(n) = aP_n(\lambda) \,,$$
$$g(n) = \frac{1}{2} P_n(\lambda) \,;$$

Case II.

$$f(n) = a\big(P_n(\lambda_1) - P_n(\lambda_2)\big) \,,$$
$$g(n) = \frac{1}{2}\big(P_n(\lambda_1) + P_n(\lambda_2)\big) \,;$$

Case III.

$$f(n) = bP_n'(\lambda) \,,$$
$$g(n) = P_n(\lambda) \,;$$

where a, b, λ, λ_1, λ_2 *are arbitrary complex numbers.*

Proof. Assume that f and g satisfy the sine equation and denote τ_f the variety generated by f. It follows from the equation that all translates of f are linear combinations of f and g, therefore the dimension of the variety τ_f is at most 2. By all means, according to spectral synthesis (see Theorem 6.7 in Section 5.6 and also [Szé04]), τ_f is generated by one or two exponential monomials. There are two different cases: if τ_f is one-dimensional, then it necessarily must be generated by an exponential function and if τ_f is two-dimensional, then it is either generated by two different exponential functions or it is generated by the functions $n \mapsto P_n(\lambda)$ and $n \mapsto P_n'(\lambda)$ with some λ in $\mathbb{C}$ (see Theorem 6.7 in Section 5.6 and also [Szé04]).

In the first case f and g are scalar multiples of a fixed exponential function:

$$f(n) = aP_n(\lambda), \qquad g(n) = bP_n(\lambda),$$

for some complex numbers a, b, λ. Substitution into (9.1) gives

$$f(n * m) = aP_n(\lambda)bP_m(\lambda) + aP_m(\lambda)bP_n(\lambda) = 2ab\,P_n(\lambda)P_m(\lambda)$$

and, on the other hand,

$$f(n * m) = a\,P_n(\lambda)P_m(\lambda),$$

thus $b = 1/2$ and a is an arbitrary complex number, hence we have Case I.

If the subspace is generated by two different exponential functions, then we have

$$f(n) = aP_n(\lambda_1) + bP_n(\lambda_2), \qquad g(n) = cP_n(\lambda_1) + dP_n(\lambda_2),$$

with some complex numbers a, b, c, d and $\lambda_1 \neq \lambda_2$. In this case $f(0) = a + b$, $g(0) = c + d$ and substituting $m = 0$ into (9.1) we have

$$f(n) = f(n)(c + d) + (a + b)g(n)$$

and

$$[aP_n(\lambda_1) + bP_n(\lambda_2)](1 - (c + d)) = (a + b)[cP_n(\lambda_1) + dP_n(\lambda_2)].$$

Since $n \mapsto P_n(\lambda_1)$ and $n \mapsto P_n(\lambda_2)$ are linearly independent, their coefficients on the left and on the right side must be equal: $a - ac - ad = ac + bc$ thus $ad + bc = a(1 - 2c)$ and similarly $ad + bc = b(1 - 2d)$. Now we put $m = 1$ into (9.1):

$$f(n * 1) = 2acP_n(\lambda_1)\lambda_1 + 2bdP_n(\lambda_2)\lambda_2 + (ad + bc)[P_n(\lambda_2)\lambda_1 + P_n(\lambda_1)\lambda_2]$$

$$= P_n(\lambda_1)[2ac\lambda_1 + (ad + bc)\lambda_2] + P_n(\lambda_2)[2bd\lambda_2 + (ad + bc)\lambda_1]$$

and, on the other hand, using the exponential property of $n \mapsto P_n$:

$$f(n * 1) = aP_{n*1}(\lambda_1) + bP_{n*1}(\lambda_2) = aP_n(\lambda_1)\lambda_1 + bP_n(\lambda_2)\lambda_2\,.$$

Here, for simplicity, $P_{n*1}(\lambda_1)$ denotes the function $n \mapsto \varphi(n * 1)$, where $\varphi(n) = P_n(\lambda_1)$. Comparing this with the previous equations we get

$$2ac\lambda_1 + (ad + bc)\lambda_2 = a\lambda_1\,,$$

$$2bd\lambda_2 + (ad + bc)\lambda_1 = b\lambda_2\,,$$

and using the relation $ad + bc = a(1 - 2c) = b(1 - 2d)$ it follows

$$a(1 - 2c)(\lambda_2 - \lambda_1) = 0$$

and

$$b(1 - 2d)(\lambda_2 - \lambda_1) = 0\,.$$

Since $\lambda_1 \neq \lambda_2$, therefore $c = d = 1/2$. Putting $m = n = 0$ into (9.1) we have $(a + b) = 2(a + b)(c + d)$, that is, $b = -a$. Consequently

$$f(n) = a(P_n(\lambda_1) - P_n(\lambda_2))\,, \qquad g(n) = \frac{1}{2}(P_n(\lambda_1) + P_n(\lambda_2))\,,$$

which is Case II.

Finally, if the variety is generated by $P_n(\lambda)$ and $P_n'(\lambda)$ with some complex λ, then we have

$$f(n) = aP_n(\lambda) + bP_n'(\lambda)\,, \qquad g(n) = cP_n(\lambda) + dP_n'(\lambda)\,,$$

where a, b, c, d are complex numbers. It is obvious that $f(0) = a$, $g(0) = c$, further $f(1) = a\lambda + b$, $g(1) = c\lambda + d$ and

$$f(n * 0) = f(n)g(0) + f(0)g(n),$$

that is

$$f(n)(1 - c) = ag(n),$$

hence

$$[aP_n(\lambda) + bP_n'(\lambda)](1 - c) = acP_n(\lambda) + adP_n'(\lambda).$$

In this case $a = 2ac$ and $b - bc = ad$ and substituting $m = 1$ into (9.1) we get

$$f(n * 1) = f(n)g(1) + f(1)g(n)$$

$$= [aP_n(\lambda) + bP_n'(\lambda)](c\lambda + d) + (a\lambda + b)[cP_n(\lambda) + dP_n'(\lambda)]$$

$$= P_n(\lambda)[a(c\lambda + d) + c(a\lambda + b)] + P_n'(\lambda)[b(c\lambda + d) + d(a\lambda + b)].$$

On the other hand

$$f(n * 1) = aP_{n*1}(\lambda) + b(P_{n*1}(\lambda))'$$

$$= aP_n(\lambda)\lambda + bP_n'(\lambda)\lambda + P_n(\lambda)b = P_n(\lambda)(a\lambda + b) + b\lambda P_n'(\lambda),$$

thus

$$a\lambda + b = 2ac\lambda + ad + cb$$

and

$$b\lambda = bc\lambda + ad\lambda + 2bd\,.$$

Substituting $m = 0$ into (9.1) we have either $a = 0$ or $c = 1/2$ and $b = 2ad$.

In the first case $b = cb$ and $b \neq 0$, hence $c = 1$, then $b\lambda = b\lambda + 2bd$ and $d = 0$.

In the second case by $b\lambda = b\lambda + 2bd$ we have either $b = d = 0$ which is covered by Case I., or $a = b = 0$, which is impossible. Consequently it follows

$$f(n) = bP_n'(\lambda), \qquad g(n) = P_n(\lambda)\,.$$

Conversely, it is easy to see that the functions given in Cases I.–III. satisfy equation (9.1), hence the theorem is proved. □

9.2   The cosine functional equation on polynomial hypergroups

The following theorem describes the general solution of the cosine equation on polynomial hypergroups.

Theorem 9.2. *Let* $(\mathbb{N}, *)$ *be the polynomial hypergroup associated with the sequence of polynomials* $(P_n)_{n \in \mathbb{N}}$ *and let* $f, g : \mathbb{N} \to \mathbb{C}$ *be functions such that* f *is not identically zero. Then* f *and* g *satisfy the cosine functional equation, that is,*

$$f(n * m) = f(n)f(m) - g(m)g(n) \tag{9.2}$$

holds for all m, n *in* $\mathbb{N}$ *if and only if* f *and* g *can be written in one of the following forms:*

Case I.

$$f(n) = aP_n(\lambda),$$
$$g(n) = \sqrt{a^2 - a}\, P_n(\lambda);$$

Case II.

$$f(n) = aP_n(\lambda_1) + (1 - a)P_n(\lambda_2),$$
$$g(n) = \sqrt{a^2 - a}\big(P_n(\lambda_1) - P_n(\lambda_2)\big);$$

Case III.

$$f(n) = P_n(\lambda) + b\,P_n'(\lambda),$$
$$g(n) = \pm b\,P_n(\lambda);$$

where a, b, λ, λ_1, λ_2 are arbitrary complex numbers.

Proof. The method of proof of Theorem 9.1 can be used.

In the first case f and g have the following form:

$$f(n) = aP_n(\lambda), \qquad g(n) = bP_n(\lambda)$$

for some complex numbers a, b, λ, hence

$$f(n * m) = a^2 P_n(\lambda)P_m(\lambda) - b^2 P_m(\lambda)P_n(\lambda) = (a^2 - b^2)P_n(\lambda)P_m(\lambda)$$

and

$$f(n * m) = aP_n(\lambda)P_m(\lambda).$$

This means that $a = a^2 - b^2$ and we have the representation given in Case I.

If the subspace τf is generated by two different exponential functions, then we have

$$f(n) = aP_n(\lambda_1) + bP_n(\lambda_2), \qquad g(n) = cP_n(\lambda_1) + dP_n(\lambda_2)$$

with some complex numbers a, b, c, d and $\lambda_1 \neq \lambda_2$. It is clear that $f(0) = a + b$ and $g(0) = c + d$. Substituting $m = 0$ into (9.2) we get

$$aP_n(\lambda_1) + bP_n(\lambda_2) = (a+b)(aP_n(\lambda_1) + bP_n(\lambda_2)) - (c+d)(cP_n(\lambda_1) + dP_n(\lambda_2))$$

$$= (a^2 - c^2)P_n(\lambda_1)P_n(\lambda_2) + (ab - cd).$$

Here $n \mapsto P_n(\lambda_1)$ and $n \mapsto P_n(\lambda_2)$ are linearly independent, thus

$$a - a^2 + c^2 = b - b^2 + d^2 = ab - cd. \tag{9.3}$$

If we put $m = 1$ into (9.2), then we have

$$f(n * 1) = [aP_n(\lambda_1) + bP_n(\lambda_2)](a\lambda_1 + b\lambda_2) - [cP_n(\lambda_1) + dP_n(\lambda_2)](c\lambda_1 + d\lambda_2)$$

and

$$f(n * 1) = aP_{n*1}(\lambda_2) + bP_{n*1}(\lambda_2) = aP_n(\lambda_1) + bP_n(\lambda_2).$$

This means that

$$(a - a^2 + c^2)\lambda_1 = (ab - cd)\lambda_2,$$

further

$$(b - b^2 + d^2)\lambda_2 = (ab - cd)\lambda_1$$

and by (9.3), we get $ab - cd = 0$, $c = \sqrt{a^2 - a}$, $b = \sqrt{b^2 - b}$ and

$$ab = cd = \sqrt{a^2 - a}\,\sqrt{b^2 - b}.$$

Taking squares we obtain $a + b = 1$, because $a = 0$ or $b = 0$ are covered by Case I. Hence it follows

$$f(n) = aP_n(\lambda_1) + (1 - a)P_n(\lambda_2), \qquad g(n) = \sqrt{a^2 - a}\big(P_n(\lambda_1) - P_n(\lambda_2)\big)$$

and this is the representation given in Case II.

Now we suppose that the variety is generated by $n \mapsto P_n(\lambda)$ and $n \mapsto P'_n(\lambda)$ with some complex λ, then we have

$$f(n) = aP_n(\lambda) + bP'_n(\lambda), \qquad g(n) = cP_n(\lambda) + dP'_n(\lambda).$$

It is obvious that $f(0) = a$, $g(0) = c$, $f(1) = a\lambda + b$, $g(1) = c\lambda + d$ and by the substitution $m = 0$ into (9.2) we get

$$f(n) = aP_n(\lambda) + bP'_n(\lambda) = a[aP_n(\lambda) + bP'_n(\lambda)] - c[cP_n(\lambda) + dP'_n(\lambda)]$$

$$= (a^2 - c^2)P_n(\lambda) + (ab - cd)P'_n(\lambda),$$

that is $a = a^2 - c^2$ and $b = ab - cd$.

Now we substitute $m = 1$ in (9.2), then we have

$$f(n * 1) = f(n)f(1) + g(1)g(n)$$

$$= [aP_n(\lambda) + bP'_n(\lambda)](a\lambda + b) + (c\lambda + d)[cP_n(\lambda) + dP'_n(\lambda)]$$

$$= (\lambda(a^2 - c^2) + ab - cd)P_n\lambda + (b^2 - d^2 + \lambda(ab - cd))P'_n(\lambda).$$

On the other hand

$$f(n * 1) = aP_{n*1}(\lambda) + b(P_{n*1}(\lambda))'$$

$$= aP_n(\lambda)\lambda + bP_n'(\lambda)\lambda + P_n(\lambda)b = P_n(\lambda)(a\lambda + b) + b\lambda P_n'(\lambda) \,,$$

thus

$$b\lambda = b^2 - d^2 + \lambda(ab - cd) \,.$$

Using that $ab - cd = b$ we get $b^2 = d^2$ and by $a = a^2 - c^2$ it follows $a = 1$ and $c = 0$. Then

$$f(n) = P_n(\lambda) + bP_n'(\lambda), \qquad g(n) = \pm b\,P_n(\lambda) \,,$$

which is the form given in Case III.

By direct substitution one verifies easily that the functions given in Cases I.–III. satisfy (9.2), hence the theorem is proved. $\qquad\square$

9.3   The Lévi–Cività functional equation

In this section we generalize the results of the previous two sections. Namely, the sine and cosine equations are particular cases of the *Lévi–Cività functional equation* which has been studied on different algebraic structures by several authors. For detailed references see [Szé91]. The Lévi–Cività functional equation (9.4) plays a basic role in the theory of finite dimensional representations of commutative groups. Here we will consider (9.4) on hypergroups with a special property. In Section 4.2 we used the term exponential family on Sturm–Liouville hypergroups. We recall this concept as follows. Let a hypergroup $(K, *)$ be given. Suppose that n is a positive integer and there exists a function $\Phi : K \times \mathbb{C}^n \to \mathbb{C}$ with the following properties:

(1) The function $x \mapsto \Phi(x, \lambda)$ is an exponential of the hypergroup K for each λ in $\mathbb{C}^n$;

(2) For every exponential m of the hypergroup K there exists a λ in $\mathbb{C}^n$ such that

$$m(x) = \Phi(x, \lambda)$$

holds for each x in K;

(3) The function $\lambda \mapsto \Phi(x, \lambda)$ is C^{∞} for each x in K.

Suppose now that every n-dimensional variety on K is generated by functions of the form

$$x \mapsto \Phi^{(j)}(x, \lambda_l),$$

for $l = 1, 2, \ldots, k$ and $j = 0, 1, \ldots, n_l - 1$, where k and $n_1, n_2, \ldots, n_k$ are positive integers such that $n_1 + n_2 + \cdots + n_k = n$. This is the case, for instance, if K is a polynomial hypergroup. We consider the Lévi–Civitá functional equation on K, that is,

$$f(x * y) = \sum_{i=1}^{n} g_i(x) h_i(y), \qquad (9.4)$$

where f, g_i and h_i are unknown complex valued functions on K for $i = 1, 2, \ldots, n$ and, in addition, we assume that the systems $\{g_1, g_2, \ldots, g_n\}$ and $\{h_1, h_2, \ldots, h_n\}$ are linearly independent. Then the n-dimensional variety τ_f contains the functions g_i and h_i for $i = 1, 2, \ldots, n$. In this case there exist $\lambda_1, \lambda, \ldots, \lambda_k$ complex numbers such that the variety τ_f is generated by the functions $x \mapsto \Phi^{(j)}(x, \lambda_l)$ for $l = 1, 2, \ldots, k$ and $j = 0, 1, \ldots, n_l - 1$, where $n_1 + n_2 + \cdots + n_k = n$. Hence we can write the unknown functions in the following form

$$f(x) = \sum_{l=1}^{k} \sum_{j=0}^{n_l-1} F_{l_j} \Phi^{(j)}(x, \lambda_l), \qquad (9.5)$$

$$g_i(x) = \sum_{l=1}^{k} \sum_{j=0}^{n_l-1} G_{l_j}^{i} \Phi^{(j)}(x, \lambda_l), \quad h_i(x) = \sum_{l=1}^{k} \sum_{j=0}^{n_l-1} H_{l_j}^{i} \Phi^{(j)}(x, \lambda_l)$$

with some complex coefficients. Denote G and H the $n \times n$ type matrices containing the coefficients of the functions g_i and h_i: if t is in $\{1, 2, \ldots, n\}$ and $t = n_1 + n_2 + \cdots + n_{l-1} + (j + 1)$, then we have

$$G_{it} = G_{l_j}^{i}, \quad H_{it} = H_{l_j}^{i}. \qquad (9.6)$$

Theorem 9.3. *The functions* $f, g_i, h_i : K \to \mathbb{C}$ *satisfy the functional equation* (9.4) *if and only if*

$$H^{\top} \cdot G = F,$$

where G and H are the matrices in (9.6) and F is the following matrix:

$$F = \begin{pmatrix} B^1 & \dots & \dots & \dots \\ \dots & B^2 & \dots & \dots \\ \dots & \dots & \dots & \dots \\ \dots & \dots & \dots & B^k \end{pmatrix},$$

where the B^l block matrix has the form

$$B^l_{ts} = \begin{cases} \binom{(t-1)+(s-1)}{s-1} F_{l_{(t-1)}}, & \text{if } (t-1)+(s-1) < n_l \\ 0, & \text{if } (t-1)+(s-1) \geq n_l. \end{cases}$$

Hence it is a $n_l \times n_l$ type matrix for all $l = 1, 2, \dots, k$. All the other elements of F are zero.

Proof. Suppose that the functions f, g_i and h_i for $i = 1, 2, \dots, n$ satisfy equation (9.4). Using the exponential property of Φ we get

$$f(x * y) = \sum_{l=1}^{k} \sum_{j=0}^{n_l-1} F_{l_j} \Phi^{(j)}(x * y, \lambda_l)$$

$$= \sum_{l=1}^{k} \sum_{j=0}^{n_l-1} F_{l_j} \sum_{t=0}^{j} \binom{j}{t} \Phi^{(t)}(x, \lambda_l) \Phi^{(j-t)}(y, \lambda_l).$$

On the other hand

$$\sum_{i=1}^{n} g_i(x) h_i(y) = \sum_{l=1}^{k} \sum_{j=0}^{n_l-1} \left(\sum_{i=1}^{n} \sum_{s=1}^{k} \sum_{r=0}^{n_s-1} H^i_{s_r} G^i_{l_j} \Phi^{(r)}(y, \lambda_s) \right) \Phi^{(j)}(x, \lambda_l)$$

and for fixed l and j the coefficients of $\Phi^{(j)}(x, \lambda_l)$ in the above equations must be equal. Hence

$$\sum_{s=1}^{k} \sum_{r=0}^{n_s-1} \left(\sum_{i=1}^{n} H^i_{s_r} G^i_{l_j} \right) \Phi^{(r)}(y, \lambda_s) = \sum_{t=j}^{n_l-1} \binom{t}{j} F_{l_t} \Phi^{(t-j)}(y, \lambda_l)$$

$$= \sum_{r=0}^{n_l-1-j} \binom{r+j}{j} F_{l_{(j+r)}} \Phi^{(r)}(y, \lambda_l)$$

and if $s \neq l$, then $\sum_{i=1}^{n} H^i_{s_r} G^i_{l_j} = 0$ for all $r = 0, 1, \dots, n_s - 1$. If $l = s$, then

$$\sum_{i=1}^{n} H_{l_r}^i G_{l_j}^i = \begin{cases} \binom{r+j}{j} F_{l_{(r+j)}}, & \text{if } r+j < n_l \\ 0, & \text{if } r+j \geq n_l \end{cases},$$

which is equivalent to the statement $H^\top \cdot G = F$ of the theorem. It is easy to see that the converse statement is true. □

We conclude this section by presenting some special cases of the Lévi–Cività functional equation, which can be solved by our method.

Example 9.1. Consider the functional equation

$$f(n * m) = g(n)h(m) + k(n) + l(m) \tag{9.7}$$

on a discrete polynomial hypergroup with basic polynomials $(P_n)_{n\in\mathbb{N}}$. Using the notation of (9.4) we have

$$f(n * m) = \sum_{i=1}^{3} g_i(n)h_i(m),$$

where

$$g_1 = g, \quad h_1 = h, \quad g_2 = k, \quad h_2 = 1, \quad g_3 = 1, \quad h_3 = l.$$

Assuming linearly independence the variety τ_f is 3-dimensional. As the identically 1 function is one of the generating functions, we have two cases:

(1) $n = 1 + 1 + 1$, that is, τ_f is generated by $P_n(\lambda_1)$, $P_n(\lambda_2)$ and 1, where $\lambda_1 \neq \lambda_2$ are complex numbers;

(2) $n = 2 + 1$, that is, τ_f is generated by $P_n(\lambda_2)$, $P_n'(\lambda)$ and 1, where λ_2 is some complex number.

In the first case

$$g_1(n) = G_{1_0}^1 P_n(\lambda_1) + G_{2_0}^1 P_n(\lambda_2) + G_{3_0}^1, \quad h_1(n) = H_{1_0}^1 P_n(\lambda_1) + H_{2_0}^1 P_n(\lambda_2) + H_{3_0}^1,$$

$$g_2(n) = G_{1_0}^2 P_n(\lambda_1) + G_{2_0}^2 P_n(\lambda_2) + G_{3_0}^2, \quad h_3(n) = H_{1_0}^3 P_n(\lambda_1) + H_{2_0}^3 P_n(\lambda_2) + H_{3_0}^3$$

and $H^2_{1_0} = H^2_{2_0} = G^3_{1_0} = G^3_{2_0} = 0$, $H^2_{3_0} = G^3_{3_0} = 1$. Applying Theorem 9.3
we have the following conditions:

$$
\begin{pmatrix} F_{1_0} & 0 & 0 \\ 0 & F_{2_0} & 0 \\ 0 & 0 & F_{3_0} \end{pmatrix} = \begin{pmatrix} H^1_{1_0} & 0 & H^3_{1_0} \\ H^1_{2_0} & 0 & H^3_{2_0} \\ H^1_{3_0} & 1 & H^3_{3_0} \end{pmatrix} \begin{pmatrix} G^1_{1_0} & G^1_{2_0} & G^1_{3_0} \\ G^2_{1_0} & G^2_{2_0} & G^2_{3_0} \\ 0 & 0 & 1 \end{pmatrix}
$$

$$
= \begin{pmatrix} H^1_{1_0}G^1_{1_0} & H^1_{1_0}G^1_{2_0} & H^1_{1_0}G^1_{3_0} + H^3_{1_0} \\ H^1_{2_0}G^1_{1_0} & H^1_{2_0}G^1_{2_0} & H^1_{2_0}G^1_{3_0} + H^3_{2_0} \\ H^1_{3_0}G^1_{1_0} + G^2_{1_0} & H^3_{1_0}G^1_{2_0} + G^2_{2_0} & H^1_{3_0}G^1_{3_0} + G^2_{3_0} + H^3_{3_0} \end{pmatrix} .
$$

In the second case

$$
g_1(n) = G^1_{1_0}P_n(\lambda_2) + G^1_{1_1}P'_n(\lambda_2) + G^1_{2_0}, \quad h_1(n) = H^1_{1_0}P_n(\lambda_2) + H^1_{1_1}P'_n(\lambda_2) + H^1_{2_0},
$$

$$
g_2(n) = G^2_{1_0}P_n(\lambda_2) + G^2_{1_1}P'_n(\lambda_2) + G^2_{2_0}, \quad h_3(n) = H^3_{1_0}P_n(\lambda_2) + H^3_{1_1}P'_n(\lambda_2) + H^3_{2_0}
$$

and $H^2_{1_0} = H^2_{1_1} = G^3_{1_0} = G^3_{1_1} = 0$, $H^2_{3_0} = G^3_{3_0} = 1$, further

$$
\begin{pmatrix} F_{1_0} & F_{1_1} & 0 \\ F_{1_1} & 0 & 0 \\ 0 & 0 & F_{2_0} \end{pmatrix} = \begin{pmatrix} H^1_{1_0}G^1_{1_0} & H^1_{1_0}G^1_{1_1} & H^1_{1_0}G^1_{2_0} + H^3_{1_0} \\ H^1_{1_1}G^1_{1_0} & H^1_{1_1}G^1_{1_1} & H^1_{1_1}G^1_{2_0} + H^3_{1_1} \\ H^1_{2_0}G^1_{1_0} + G^2_{1_0} & H^3_{1_0}G^1_{1_1} + G^2_{1_1} & H^1_{2_0}G^1_{2_0} + G^2_{2_0} + H^3_{2_0} \end{pmatrix} .
$$

Comparing these equations, we get the following solutions of (9.7), where
the role of $\{g, k\}$ and $\{h, l\}$ can be interchanged:

$$
f(n) = G^1_{3_0}H^1_{3_0} + G^2_{3_0} + H^3_{3_0} ,
$$

$$
g(n) = G^1_{1_0}P_n(\lambda_1) + G^1_{2_0}P_n(\lambda_2) + G^1_{3_0} ,
$$

$$
h(n) = H^1_{3_0} ,
$$

$$k(n) = (-G^1_{1_0}H^1_{3_0})P_n(\lambda_1) + (-G^1_{2_0}H^1_{3_0})P_n(\lambda_2) + G^2_{3_0} \,,$$

$$l(n) = H^3_{3_0}$$

and

$$f(n) = G^1_{1_0}H^1_{1_0}P_n(\lambda) + (G^1_{2_0}H^1_{2_0} + H^3_{2_0} + G^2_{2_0}) \,,$$

$$g(n) = G^1_{1_0}P_n(\lambda) + G^1_{1_1}P'_n(\lambda) + G^1_{2_0} \,,$$

$$h(n) = H^1_{1_0}P_n(\lambda) + H^1_{2_0} \,,$$

$$k(n) = (-G^1_{1_0}H^1_{2_0})P_n(\lambda) + (-G^1_{1_1}H^1_{2_0})P'_n(\lambda) + G^2_{2_0} \,,$$

$$l(n) = (-G^1_{2_0}H^1_{1_0})P_n(\lambda) + H^3_{2_0} \,.$$

Chapter 10

Difference equations on polynomial hypergroups

10.1 Introduction

A linear difference equation with nonconstant coefficients has the following general form:

$$a_N(n)f(n+N)+a_{N-1}(n)f(n+N-1)+\cdots+a_1(n)f(n+1)+a_0(n)f(n) = g(n)\,,$$

where the functions $a_0, a_1, \ldots, a_N, g : \mathbb{N} \to \mathbb{C}$ are given with a_N is not identically zero and N, k are fixed nonnegative integers. The above equation is supposed to hold for some unknown function $f : \mathbb{N} \to \mathbb{C}$ or $f : \mathbb{Z} \to \mathbb{C}$ and for each n in $\mathbb{N}$ or for each n in $\mathbb{Z}$, depending on the nature of the problem. In what follows we shall prefer the case $f : \mathbb{N} \to \mathbb{C}$.

By the classical theory of differential equations the solution space of the above equation can be described completely in the *constant coefficient case*, that is, if the functions $a_0, a_1, \ldots, a_N$ are constants. In this case the solution space is generated by *exponential monomial solutions*, which arise from the roots of the characteristic polynomial, called *characteristic roots*.

Much less is known in the case of nonconstant coefficients. In this section we offer a method to solve some types of homogeneous linear difference equations with nonconstant coefficients by transforming these equations into homogeneous linear difference equations with constant coefficients over *hypergroups*. This method is based on some theory of homogeneous linear difference equations with constant coefficients on hypergroups developed along the lines of the classical theory over $\mathbb{N}$. The basic idea is that the role of exponential functions is played by the generating polynomials of some polynomial hypergroups.

Most of the results of this chapter are are taken from [Oro06a]. We remark that the major part of this work can be generalized to the case of *signed hypergroups*, as they are presented in [Ros98]. Nevertheless, in the forthcoming pages we restrict ourselves to polynomial hypergroups in the sense of [BH95]. For other results concerning difference equations on hypergroups see [EL07].

10.2 Difference equations with 1-translation

In the classical theory of difference equations the translate of a function by n and the translation of the function n-times by 1 give the same result for all n in $\mathbb{N}$. But in the hypergroup case there are two different ways to define difference equations along these two interpretations. In this section we deal with the latter one. We introduce the notation

$$\mathcal{T}f(n) = \mathcal{T}_1 f(n) = f(n * 1)$$

for any $f : \mathbb{N} \to \mathbb{C}$ and n in $\mathbb{N}$, moreover $\mathcal{T}^0 f = f$ and $\mathcal{T}^N f = \mathcal{T}(\mathcal{T}^{N-1} f)$ for each integer $N > 1$. Obviously, $\mathcal{T}$ is a linear operator on the linear space $\mathbb{C}^{\mathbb{N}}$ of all complex valued functions on $\mathbb{N}$. If Q is any polynomial with complex coefficients, then $Q(\mathcal{T})$ has the obvious meaning. Let $N \geq 1$ be in $\mathbb{N}$, let $a_0, \dots, a_N$ be complex numbers and suppose that $a_N \neq 0$. We shall consider functional equations of the form

$$Q(\mathcal{T})f = a_N \mathcal{T}^N f(n) + a_{N-1} \mathcal{T}^{N-1} f(n) + \cdots + a_0 f(n) = 0, \qquad (10.1)$$

which is called a *homogeneous linear difference equation* of order N on the hypergroup K with constant coefficients associated to the polynomial Q. The polynomial Q is called the *characteristic polynomial* of (10.1) and its roots are called the *characteristic roots of* (10.1). The *solution space* of (10.1) is the kernel of the linear operator $Q(\mathcal{T})$, hence it is a linear subspace of the function space $\mathbb{C}^{\mathbb{N}}$. This solution space is translation invariant in the sense that if f is a solution, then $\mathcal{T}f$ is a solution, too.

Theorem 10.1. *If Q is a complex polynomial of degree $N \geq 1$, then the solution space of* (10.1) *has dimension N.*

Proof. Suppose that $f : \mathbb{N} \to \mathbb{C}$ is a solution of equation (10.1). Since $f(0 * 1) = f(1)$ we have for $n \geq 1$

$$f(n * 1) = \sum_{k=n-1}^{n+1} c(n, 1, k) f(k) = \alpha_n f(n-1) + \beta_n f(n) + \gamma_n f(n+1),$$

where $(\alpha_n)_{n \in \mathbb{N}}$, $(\beta_n)_{n \in \mathbb{N}}$ and $(\gamma_n)_{n \in \mathbb{N}}$ are the sequences appearing in the definition of the polynomial hypergroup. Considering equation (10.1) for $n = 0$ we get

$$a_N \gamma_{N-1} \cdots \gamma_1 \gamma_0 f(N) + \sum_{i=0}^{N-1} k_{N,i} f(i) = 0$$

with some complex numbers $k_{N,i}$ ($i = 0, \ldots, N-1$). Since obviously we have $a_N \gamma_{N-1} \cdots \gamma_0 \neq 0$, hence $f(N)$ is determined by $f(0), f(1), \ldots, f(N-1)$ and it is easy to see by induction that $f(n)$ is uniquely determined by these values. $\qquad \square$

Theorem 10.2. *If the complex number λ is a characteristic root of* (10.1) *with multiplicity m, then all the functions $n \mapsto P_n^{(k)}(\lambda)$ are solutions of equation* (10.1) *for $k = 0, 1, \ldots, m-1$.*

Proof. By the exponential property of the function $n \mapsto P_n(\lambda)$ we have

$$\mathcal{T} P_n(\lambda) = P_n(\lambda) P_1(\lambda) = \lambda P_n(\lambda), \quad \mathcal{T}^t P_n(\lambda) = \lambda^t P_n(\lambda),$$

thus we can see immediately that this function is a solution of (10.1):

$$Q(\mathcal{T}) P_n(\lambda) = (\lambda^N + a_{N-1} \lambda^{N-1} + \cdots + a_1 \lambda + a_0) P_n(\lambda) = 0.$$

In order to prove that $n \mapsto P_n^{(k)}(\lambda)$ are also solutions for $1 \leq k \leq m$ we need the translates of $P_n^{(k)}(\lambda)$. By easy calculation we get that

$$\mathcal{T}^r P_n^{(k)}(\lambda) = \sum_{t=0}^{\min(r,k)} \binom{r}{t} \lambda^{r-t} \frac{k!}{(k-t)!} P_n^{(k-t)}(\lambda)$$

for r in $\mathbb{N}$, therefore we have

$$Q(\mathcal{T})P_n^{(k)}(\lambda) = \sum_{r=0}^{N} a_r \mathcal{T}^r P_n^{(k)}(\lambda)$$

$$= \sum_{r=0}^{N} a_r \left(\sum_{t=0}^{\min(r,k)} \binom{r}{t} \lambda^{r-t} \frac{k!}{(k-t)!} P_n^{(k-t)}(\lambda) \right)$$

$$= \sum_{t=0}^{k} \binom{k}{t} \left(\sum_{r=t}^{N} a_r \frac{r!}{(r-t)!} \lambda^{r-t} \right) P_n^{(k-t)}(\lambda) = 0 \,,$$

as the t-th derivative of the characteristic polynomial at λ is equal to zero for $0 \le t \le m-1$. $\square$

Lemma 7. *Let k be a positive integer and $l_1, \ldots, l_k$ nonnegative integers. If $\lambda_1, \ldots, \lambda_k$ are different complex numbers, then the functions $P_n^{(i)}(\lambda_j)$ are linearly independent for $j = 1, 2, \ldots, k$ and $i = 0, 1, \ldots, l_j$.*

Proof. First we show that $P_n(\lambda_1), \ldots, P_n(\lambda_k)$ are linearly independent if $\lambda_1, \ldots, \lambda_k$ are different. If it is not the case, then there are complex numbers $a_1, a_2, \ldots, a_k$, not all equal to zero, with the property $\sum_{i=1}^{k} a_i P_n(\lambda_i) = 0$, which contradicts to the fact that for some constant C the following equation holds

$$\begin{vmatrix} P_0(\lambda_1) & \ldots & P_0(\lambda_k) \\ P_1(\lambda_1) & \ldots & P_1(\lambda_k) \\ & \ddots & \\ P_k(\lambda_1) & \ldots & P_k(\lambda_k) \end{vmatrix} = C \begin{vmatrix} 1 & \ldots & 1 \\ \lambda_1 & \ldots & \lambda_k \\ & \ddots & \\ \lambda_1^k & \ldots & \lambda_k^k \end{vmatrix} \ne 0 \,.$$

Now assume that there exist $\lambda_1, \ldots, \lambda_k$ different complex numbers such that the functions $n \mapsto P_n^{(i)}(\lambda_j)$ are linearly dependent for $j = 1, 2, \ldots, k$ and $i = 0, 1, \ldots, l_j$ for some positive integers $l_1, \ldots, l_k$. Suppose that k is the minimal positive integer with this property and also suppose that $l_1 + \cdots + l_k$ is minimal. It means that there exist complex numbers $a_{j,i}$, not all equal to zero for $j = 1, \ldots, k$ and $i = 0, \ldots, l_j$ such that

$$\sum_{j=1}^{k} \sum_{i=0}^{l_j} a_{j,i} P_n^{(i)}(\lambda_j) = 0 \tag{10.2}$$

holds with $a_{j,l_j} \ne 0$. Translating equation (10.2) by 1 we have

$$\sum_{j=1}^{k}\sum_{i=1}^{l_j} a_{j,i} \left(\lambda_j P_n^{(i)}(\lambda_j) + P_n^{(i-1)}(\lambda_j) \right) + \sum_{j=1}^{k} a_{j,0} \lambda_j P_n(\lambda_j) = 0$$

and if we subtract equation (10.2) times λ_1 from this equation we get an expression which does not contain $P_n^{(l_1)}(\lambda_1)$:

$$\sum_{i=1}^{l_1-1} c_{1,i} P_n^{(i)}(\lambda_1) + \sum_{j=2}^{k} a_{j,l_j}(\lambda_1 - \lambda_j) P_n^{(l_j)} + \sum_{j=2}^{k}\sum_{i=0}^{l_j-1} c_{j,i} P_n^{(i)}(\lambda_j) = 0$$

with some constants $c_{j,i}$ and this means that either k or $l_1 + \cdots + l_k$ is not minimal. $\qquad\square$

Using Theorem 10.1, Theorem 10.2 and Lemma 7 we can completely characterize the solution space of (10.1).

Theorem 10.3. *Let Q be a complex polynomial of degree $N \geq 1$ with all different complex zeros $\lambda_1, \lambda_2, \ldots, \lambda_k$, where the multiplicity of λ_j is l_j $(j = 1, 2, \ldots, k)$. Then the function $f : \mathbb{N} \to \mathbb{C}$ is a solution of (10.1) if and only if it is the linear combination of functions of the form $n \mapsto P_n^{(i)}(\lambda_j)$ with $j = 1, 2, \ldots, k$ and $i = 0, 1, \ldots, l_j - 1$.*

10.3 Difference equations with general translation

Let us consider the following equation:

$$a_N \mathcal{T}_N f(n) + a_{N-1} \mathcal{T}_{N-1} f(n) + \cdots + a_0 f(n) = 0, \qquad (10.3)$$

where $f : \mathbb{N} \to \mathbb{C}$ is a function, N is a positive integer and $a_N, \ldots, a_0$ are complex numbers. Here $\mathcal{T}_N$ is the operator defined by

$$\mathcal{T}_N f(n) = f(n * N)$$

for each n in $\mathbb{N}$. We note that, in general, $\mathcal{T}_N \neq \mathcal{T}_1^N$ and equation (10.3) can be written in the form

$$a_N f(n * N) + a_{N-1} f(n * (N-1)) + \cdots + a_0 f(n) = 0.$$

It is easy to see that the solution space of (10.3) is a linear subspace of $\mathbb{C}^N$ with dimension N. We will show that this solution space is generated by similar functions like in the case of (8.2), but the characteristic polynomial is different: it depends on the basic generating polynomials of the hypergroup.

Theorem 10.4. *The function $f : \mathbb{N} \to \mathbb{C}$ is a solution of* (10.3) *if and only if it is the linear combination of functions of the form $n \mapsto P_n^{(i)}(\lambda_j)$ with $j = 1, 2, \ldots, k$ and $i = 0, 1, \ldots, l_j - 1$, where $\lambda_1, \lambda_2, \ldots, \lambda_k$ are different complex zeros of the polynomial*

$$\lambda \mapsto a_N P_N(\lambda) + a_{N-1} P_{N-1}(\lambda) + \cdots + a_1 P_1(\lambda) + a_0 \qquad (10.4)$$

and the multiplicity of λ_j is l_j $(j = 1, 2, \ldots, k)$.

Proof.   It is enough sufficient to show that the functions $n \mapsto P_n^{(i)}(\lambda_j)$ with $j = 1, 2, \ldots, k$ and $i = 0, 1, \ldots, l_j - 1$ are solutions. Since

$$\mathcal{T}_m(P_n^{(i)}(\lambda)) = \sum_{t=0}^{i} \binom{i}{t} P_m^{(t)}(\lambda) P_n^{(i-t)}(\lambda)$$

for all m in $\mathbb{N}$, substituting $P_n^{(i)}(\lambda)$ instead of $f(n)$ in (10.3) we get

$$\sum_{m=0}^{N} a_m \mathcal{T}_m(P_n^{(i)}(\lambda)) = \sum_{m=0}^{N} a_m \sum_{t=0}^{i} \binom{i}{t} P_m^{(t)}(\lambda) P_n^{(i-t)}(\lambda)$$

$$= \sum_{t=0}^{i} \binom{i}{t} P_n^{(i-t)}(\lambda) \left(\sum_{m=0}^{N} a_m P_m^{(t)}(\lambda) \right) = 0 \,,$$

which holds if λ is a root of (10.4) with a multiplicity higher than i. $\square$

Here we present some simple examples to illustrate our method.

Example 10.1. We consider the equation

$$\mathcal{T} f = 0 \,. \qquad (10.5)$$

On the Chebyshev hypergroup we have

$$\mathcal{T}f(n) = \frac{1}{2}\left(f(n+1) + f(|n-1|)\right),$$

hence (10.5) has the form

$$f(n+1) + f(|n-1|) = 0$$

for $n = 0, 1, \ldots$. With $n = 0$ we have $f(1) = 0$ and with $n \geq 0$ it follows $f(n+2) + f(n) = 0$, which implies $f(2n+1) = 0$ and $f(2n) = (-1)^n f(0)$. Clearly, this is a solution with an arbitrary complex number $f(0)$.

On the Legendre hypergroup we have

$$\mathcal{T}f(n) = \frac{n+1}{2n+1}f(n+1) + \frac{n}{2n+1}f(|n-1|),$$

hence (10.5) has the form

$$(n+1)f(n+1) + nf(|n-1|) = 0$$

for $n \geq 0$. With $n = 0$ we have $f(1) = 0$ and with $n \geq 0$ it follows $(n+2)f(n+2) + (n+1)f(n) = 0$, which implies $f(2n+1) = 0$, moreover $f(2n) = (-1)^n \frac{(2n-1)!!}{(2n)!!}f(0)$. Again, this is a solution with an arbitrary complex value $f(0)$.

One observes that in the first case $f(n) = f(0) \cdot \mathcal{T}_n(0)$ and in the second case $f(n) = f(0) \cdot P_n(0)$, where T_n, respectively P_n denotes the n-th Chebyshev polynomial, respectively the n-th Legendre polynomial. This is a simple consequence of our previous results. Indeed, the characteristic polynomial of equation (10.5) is $Q(\lambda) = \lambda$, hence the only characteristic root is $\lambda = 0$ with multiplicity 1. Hence, on any polynomial hypergroup with generating polynomials $(P_n)_{n \in \mathbb{N}}$, by Theorem 10.3, the general solution of the difference equation (10.5) has the form $f(n) = f(0) \cdot P_n(0)$.

Now we consider the following problem: find all solutions $f : \mathbb{N} \to \mathbb{C}$ of the difference equation

$$(n+2)f(n+2) - (2n+3)f(n+1) + (n+1)f(n) = 0 \qquad (10.6)$$

with $f(0) = f(1)$, which is a homogeneous linear difference equation with nonconstant coefficients. Observe that, by introducing

$$g(n) = (n+1)f(n+1) - (n+1)f(n)$$

for $n = 0, 1, \dots$ we have $g(n+1) - g(n) = 0$, which means that g is constant and $g(n) = g(0) = f(1) - f(0) = 0$. It follows

$$(n+1)f(n+1) - (n+1)f(n) = 0\,,$$

which implies again that f is constant: $f(n) = f(0)$ for each n in $\mathbb{N}$. Then again one can realize that (10.6) is exactly the difference equation

$$\mathcal{T}f = f \qquad (10.7)$$

on the Legendre hypergroup, which is a very special case of (10.1) and can be solved by the method we offered above. Indeed, the characteristic polynomial has the form $Q(\lambda) = \lambda - 1$ and the only characteristic root is $\lambda = 1$ with multiplicity 1. According to Theorem 10.3, the general solution of the difference equation (10.6) has the form $f(n) = f(0) \cdot P_n(1)$, where P_n is the n-th Legendre polynomial. As $P_n(1) = 1$ for each n in $\mathbb{N}$, we have that all solutions of (10.6) satisfying $f(0) = f(1)$ are constant.

We can modify equation (10.7) to consider

$$\mathcal{T}f = c \cdot f\,, \qquad (10.8)$$

where c is a complex parameter. This is the eigenvalue problem for the translation operator $\mathcal{T}$ on any polynomial hypergroup with generating polynomials $(P_n)_{n \in \mathbb{N}}$. In this case the characteristic polynomial is $Q(\lambda) = \lambda - c$ having the only characteristic root $\lambda = c$ with multiplicity 1. Hence each complex number c is an eigenvalue with the corresponding eigenfunction $n \mapsto P_n(c)$.

Example 10.2. The study of higher order difference equations leads naturally to the study of generalized polynomial functions on polynomial hypergroups. Here we work out a simple special case, only. Consider the difference equation

$$\mathcal{T}^2 f - 2\mathcal{T} f + f = 0 \,, \tag{10.9}$$

or $\Delta^2 f = 0$, where we use the notation $\Delta = \mathcal{T} - I$ and I is the identity operator. The generating polynomials of the underlying polynomial hypergroup are the polynomials $(P_n)_{n \in \mathbb{N}}$. The characteristic polynomial of (10.9) is $Q(\lambda) = \lambda^2 - 2\lambda + 1 = (\lambda - 1)^2$, hence the only characteristic root is $\lambda = 1$ with multiplicity 2. By Theorem 10.3 the general solution of (10.9) has the form

$$f(n) = A \cdot P_n(1) + B \cdot P_n'(1) = A + B \cdot P_n'(1)$$

with arbitrary complex constants A, B. We know from the previous results in Section 2.1 that $n \mapsto B \cdot P_n'(1)$ represents a general additive function on the given hypergroup, hence equation (10.9) can be considered as a characterization of affine functions on polynomial hypergroups – exactly as in the group case.

Example 10.3. Our last example illustrates the application of Theorem 10.4. We consider the difference equation

$$\mathcal{T}_2 f - 2\mathcal{T}_1 f + f = 0 \tag{10.10}$$

on the polynomial hypergroup generated by the sequences $(\alpha_n)_{n \in \mathbb{N}}$, $(\beta_n)_{n \in \mathbb{N}}$ and $(\gamma_n)_{n \in \mathbb{N}}$. By Theorem 10.4 the general solution of (10.10) can be described with the help of the roots of the polynomial

$$\lambda \mapsto P_2(\lambda) - 2P_1(\lambda) + 1 \,,$$

where P_n denotes the n-th basic polynomial for all n in $\mathbb{N}$. Using the recursive formula for $n = 1$ and the property $\alpha_n + \beta_n + \gamma_n = 1$ we get

$$\gamma_1(P_2(\lambda) - 2P_1(\lambda) + 1) = (\lambda - 1)(\lambda - (\gamma_1 - \alpha_1)),$$

hence the solutions are the functions

$$f(n) = AP_n(1) + BP_n(\gamma_1 - \alpha_1) = A + BP_n(\gamma_1 - \alpha_1) \qquad (10.11)$$

with arbitrary complex numbers A, B. On the Chebyshev hypergroup, where the recursive formula for the Chebyshev polynomials $(T_n)_{n \in \mathbb{N}}$ is

$$\lambda T_n(\lambda) = \frac{1}{2}T_{n+1}(\lambda) + \frac{1}{2}T_{|n-1|}(\lambda)$$

with $T_0(\lambda) = 1$ and $T_1(\lambda) = \lambda$, the above equation (10.10) has the form

$$f(n + 4) - 2f(n + 3) + 2f(n + 2) - 2f(n + 1) + f(n) = 0$$

with the initial conditions

$$f(2) = 2f(1) - f(0) \qquad \text{and} \qquad f(3) = f(1).$$

The general solution has the form

$$f(n) = A + B \cdot T_n(0)$$

with arbitrary complex numbers A, B and, more explicitly, we can write $f(2n) = A + B(-1)^n$ and $f(2n + 1) = 0$ for each n in $\mathbb{N}$. As in this case the problem reduces to a linear homogeneous difference equation with constant coefficients, the same result can be derived from the classical theory. Nevertheless, in the case of the Legendre hypergroup one obtains a linear homogeneous difference equation with nonconstant coefficients and the classical methods cannot be directly applied, but, by the virtue of (10.11), we know that the solutions are the functions

$$f(n) = A + B \cdot P_n\left(\frac{1}{3}\right),$$

where P_n is the n-th Legendre polynomial. Similarly, in the case of the Chebyshev hypergroup of the second kind the recursive formula has the form

$$\lambda U_n(\lambda) = \frac{n+2}{2n+2} U_{n+1}(\lambda) + \frac{n}{2n+2} U_{|n-1|}(\lambda)$$

with $U_0(\lambda) = 1$ and $U_1(\lambda) = \lambda$, thus the solutions of (10.10) are the functions

$$f(n) = A + B \cdot U_n\left(\frac{1}{2}\right).$$

Chapter 11

Stability problems on hypergroups

11.1 Stability of exponential functions on hypergroups

Stability theory of functional equations on groups, on semigroups and on different algebraic structures has attracted the attention of several mathematicians recently. In this respect see [Szé00]. According to our knowledge stability problems for functional equations on hypergroups have not been considered so far. In this section we show that using similar ideas as in the case of groups, semigroups, etc., analogous results can be obtained.

First we deal with the stability of exponential functions on hypergroups. The following result is similar to that of [Szé82b].

Theorem 11.1. *Let K be a hypergroup and let $f, g : K \to \mathbb{C}$ be continuous functions with the property that the function $y \mapsto \int_K f\, d(\delta_x * \delta_y) - f(x)g(y)$ is bounded for all y in K. Then either f is bounded, or g is exponential.*

Proof. It is easy to see ([BH95]) that for any continuous function $\varphi : K \to \mathbb{C}$ and for any y in K the functions $x \mapsto \int_K \varphi\, d(\delta_x * \delta_y)$ and $y \mapsto \int_K \varphi\, d(\delta_x * \delta_y)$ are continuous. First of all we can check easily that the associativity of the convolution implies that

$$\int_K \int_K \varphi(s)\, d(\delta_t * \delta_z)(s)\, d(\delta_x * \delta_y)(t) = \int_K \int_K \varphi(s)\, d(\delta_x * \delta_t)(s)\, d(\delta_y * \delta_z)(t)$$

holds for all x, y, z in K. Actually, on the left hand side we have

$$\int_K \varphi\, d[(\delta_x * \delta_y) * \delta_z]$$

and on the right hand side

161

$$\int_K \varphi \, d[\delta_x * (\delta_y * \delta_z)] \, .$$

The assumption of the theorem means that

$$\left| \int_K f(s) \, d(\delta_t * \delta_z) - f(t)g(z) \right| \le M(z)$$

holds for all x, y in K with some given continuous function $M : K \to \mathbb{C}$. Integrating with respect to the measure $\delta_x * \delta_y$ and with respect to the variable t we have for all x, y, z in K:

$$\left| \int_K \int_K f(s) \, d(\delta_t * \delta_z)(s) \, d(\delta_x * \delta_y)(t) - \int_K f(t)g(z) \, d(\delta_x * \delta_y)(t) \right| \le M(z) \, .$$

Let

$$A(x, y, z) = \int_K \int_K f(s) \, d(\delta_t * \delta_z)(s) \, d(\delta_x * \delta_y)(t) - \int_K f(t)g(z) \, d(\delta_x * \delta_y)(t)$$

for all x, y, z.

Using again the assumption of the theorem we have

$$\left| \int_K f(s) \, d(\delta_x * \delta_t)(s) - f(x)g(t) \right| \le M(t)$$

for all x, t in K. Now we integrate both sides with respect to the measure $\delta_y * \delta_z$ and with respect to the variable t to get

$$\left| \int_K \int_K f(s) \, d(\delta_x * \delta_t)(s) \, d(\delta_y * \delta_z)(t) - \int_K f(x)g(t) \, d(\delta_y * \delta_z)(t) \right|$$

$$\le \int_K M(t) \, d(\delta_y * \delta_z)(t)$$

for all x, y, z in K. Let

$$B(x, y, z) = \int_K \int_K f(s) \, d(\delta_x * \delta_t)(s) \, d(\delta_y * \delta_z)(t) - \int_K f(x)g(t) \, d(\delta_y * \delta_z)(t)$$

for all x, y, z in K.

Our next observation is that

$$\left| f(x)g(y)g(z) - \int_K f(t)\,d(\delta_x * \delta_y)(t)g(z) \right| \leq g(z)M(y)$$

holds for any x, y, z in K. Let

$$C(x, y, z) = f(x)g(y)g(z) \int_K f(t)\,d(\delta_x * \delta_y)(t)g(z)$$

for all x, y, z in K.

Finally, we consider the identity

$$A(x, y, z) - B(x, y, z) - C(x, y, z) = f(x)\left(\int_K g(t)\,d(\delta_y * \delta_z)(t) - g(y)g(z) \right),$$

which holds for all x, y, z in K. From the above inequalities we can see that the left hand side is bounded for any fixed y, z in K, hence if g is not an exponential, then f is bounded and the theorem is proved. $\square$

11.2 Stability of additive functions on hypergroups

The stability of additive functions on hypergroups can be derived easily in the presence of an invariant mean, which depends on the hypergroup. However, an invariant mean always exists on commutative hypergroups, by the Markov–Kakutani fixed point theorem (see [Lau83]), as we have seen in Section 1.3, Theorem 1.4.

Theorem 11.2. *Let K be a discrete amenable hypergroup and $f : K \to \mathbb{C}$ a function with the property that the function*

$$(x, y) \mapsto \int_K f\,d(\delta_x * \delta_y) - f(x) - f(y)$$

is bounded on $K \times K$. Then there exists an additive function $a : K \to \mathbb{C}$ such that $f - a$ is bounded.

Proof. We can prove this theorem following the lines of Remark 17 in [Szé85]. For any fixed y in K we define

$$a(y) = M_x\left[\int_K f\,d(\delta_x * \delta_y) - f(x) \right],$$

where M is any invariant mean on K and M_x indicates that we apply M to the (obviously bounded) function in the brackets, as a function of x. Now we have

$$\int_K a(t)\, d(\delta_y * \delta_z)(t) - a(y) - a(z)$$

$$= M_x \left[\int_K \int_K f(s)\, d(\delta_z * \delta_t)(s)\, d(\delta_y * \delta_x)(t) - \int_K f(t)\, d(\delta_x * \delta_y)(t) \right]$$

$$- M_x \left[\int_K f(t)\, d(\delta_x * \delta_z)(t) - f(x) \right]$$

for all y, z in K. If $\varphi : K \to \mathbb{C}$ is the function defined by

$$\varphi(x) = \int_K f(t)\, d(\delta_x * \delta_z)(t) - f(x)$$

for all x, z in K, then the above equation can be written in the form

$$\int_K a(t)\, d(\delta_y * \delta_z)(t) - a(y) - a(z) = M_x \left[\varphi(t)\, d(\delta_x * \delta_y)(t) \right] - M(\varphi)$$

and the right hand side is 0 by the invariance of M. It means that a is additive. On the other hand,

$$|a(y) - f(y)| = \left| M_x \left[\int_k f(t)\, d(\delta_x * \delta_y)(t) - f(x) - f(y) \right] \right| \leq C$$

holds for all y in K, where C is a bound for the function

$$(x, y) \mapsto \int_K f\, d(\delta_x * \delta_y) - f(x) - f(y)\,.$$

The theorem is proved.   □

11.3 Superstability of a mixed-type functional equation

In this section we apply the results of the previous two sections in order to illustrate of the methods presented above.

Theorem 11.3. *Let K be a discrete amenable hypergroup and let $f, g, h : K \to \mathbb{C}$ be functions. The function*

$$(x, y) \mapsto f(x * y) - f(x)g(y) - h(y) \tag{11.1}$$

is bounded on $K \times K$ if and only if one of the following possibilities holds:

(1) f is identically zero, g is arbitrary and h is bounded,

(2) f is constant and $f \cdot g + h$ is bounded,

(3) f, g, h are bounded,

(4) $f = a + b_1$, $g = 1$, $h = a + b_2$,

(5) $f = cm + \varphi + b_1$, $g = m$, $h = \varphi + b_2$,

where $a : K \to \mathbb{C}$ is additive, $m : K \to \mathbb{C}$ is exponential, $\varphi : K \to \mathbb{C}$ satisfies

$$\varphi(x * y) = \varphi(x)m(y) + \varphi(y) \tag{11.2}$$

for each x, y in K, $b_1, b_2 : K \to \mathbb{C}$ are bounded, c, d are constants and $b_1 \cdot m$ is bounded.

Proof. As the first three cases are obvious we may suppose that f is unbounded. Then, by Theorem 11.1, the function $g = m$ is an exponential. First we assume that $m = 1$. Then the function

$$(x, y) \mapsto f(x * y) - f(x) - h(y)$$

is bounded. The substitution $x = e$ gives that $f - h$ is bounded and we infer that the function

$$(x, y) \mapsto f(x * y) - f(x) - f(y)$$

is bounded, which implies, by Theorem 11.2, that there exist bounded functions $b_1, b_2 : K \to \mathbb{C}$ such that we have Case (4).

Now we assume that $m \neq 1$. For x, y in K we have with the notation

$$F(x, y) = f(x * y) - f(x)m(y) - h(y)$$

that

$$h(y * z) - h(y)m(z) - h(z) = \left[f(x * y * z) - f(x * y)m(z) - h(z) \right]$$

$$+ \left[f(x * y) - f(x)m(y) - h(y) \right] m(z) - \left[f(x * y * z) - f(x)m(y)m(z) - h(y * z) \right]$$

$$= F(x * y, z) + F(x, y)m(z) - F(x, y * z) \,,$$

hence it is bounded on $K \times K$. Let M be an invariant mean on K and we define

$$\varphi(y) = M_x[h(x * y) - h(x)m(y)]$$

for arbitrary y in K. Then we have for each y, z in K

$$\varphi(y * z) - \varphi(y)m(z) - \varphi(z)$$

$$= M_x[h(x * y * z) - h(x)m(y * z) - h(x * y)m(z) + h(x)m(y)m(z)]$$

$$- M_x[h(x * z) - h(x)m(z)]$$

$$= M_x[h(x * y * z) - h(x)m(y)m(z) - h(x * y)m(z) + h(x)m(y)m(z)]$$

$$- M_x[h(x * y * z) - h(x * y)m(z)] = 0 \,,$$

that is, φ satisfies (11.2). Moreover, we have for each y in K

$$\varphi(y) - h(y) = M_x[h(x * y) - h(x)m(y) - h(y)] \,,$$

which is bounded. Finally, from the condition of the theorem by the substitution $x = e$ it follows that the function

$$x \mapsto f(x) - f(e)m(x) - \varphi(x)$$

is bounded. But in this case this bounded function must be identically zero, as it is easy to check, hence we arrive at the last case listed above.

The converse statement can be verified by a straightforward calculation. The theorem is proved. □

We note that the condition on $b_1 \cdot m$ obviously means that if $b_1 = 0$, then m is an arbitrary exponential, while in the case $b_1 \neq 0$ the exponential m must be bounded. Actually we have proved the following theorem.

Theorem 11.4. *Let K be a discrete amenable hypergroup and suppose that for the unbounded functions $f, g : K \to \mathbb{C}$ the function*

$$(x, y) \mapsto f(x * y) - f(x)g(y) - f(y)$$

is bounded. Then g is an exponential and f, g satisfy the functional equation

$$f(x * y) = f(x)g(y) + f(y) \tag{11.3}$$

for each x, y in K.

In the case of commutative hypergroups we can go one step further.

Theorem 11.5. *Let K be a discrete commutative hypergroup and suppose that for the unbounded functions $f, g : K \to \mathbb{C}$ the function*

$$(x, y) \mapsto f(x * y) - f(x)g(y) - f(y)$$

is bounded. Then g is an exponential and $f = c \cdot m - c$ with some constant c.

Proof. Let K be commutative. Interchanging x and y in (11.3) we obtain $f(x)(g(y) - 1) = f(y)(g(x) - 1)$, which, by $g \neq 1$, implies our statement. □

In the commutative case we can formulate another theorem.

Theorem 11.6. *Let K be a discrete commutative hypergroup and suppose that for the functions $f, g, h : K \to \mathbb{C}$ the function*

$$(x, y) \mapsto f(x * y) - f(x)g(y) - h(y) \tag{11.4}$$

is bounded. If f is unbounded, then f, g, h have any of the following forms

(1) $f(x) = a(x) + c, g(x) = 1, h(x) = a(x)$,

(2) $f(x) = (c + d)m(x) - d, g(x) = m(x), h(x) = dm(x) - d + b(x)$,

where $a : K \to \mathbb{C}$ *is additive,* $m : K \to \mathbb{C}$ *is exponential,* $b : K \to \mathbb{C}$ *is bounded and* c, d *are constants. Conversely, with this choice the function given in* (11.4) *is bounded.*

Proof. We have to show only that if the hypergroup K is commutative in the previous theorem, then the solution of (11.3) is represented by the two possibilities given here. If $g = 1$, then (11.3) reduces to

$$f(x * y) = f(x) + h(y)$$

for x, y in K. Interchanging x and y we get that $f = h + c$ with some constant c, further

$$h(x * y) = h(x) + h(y),$$

that is $h = a$ is additive and we have case (1) above.

Suppose that $g \neq 1$. By Theorems 11.3 and 11.4 we have that $g = m$ is an exponential and the functions f, h have the form given in case (5) in Theorem 11.3, where

$$\varphi(x * y) = \varphi(x)m(y) + \varphi(y)$$

for x, y in K. Interchanging x and y we get

$$\varphi(x)\big(m(y) - 1\big) = \varphi(y)\big(m(x) - 1\big)$$

and, by $m \neq 1$ it follows $\varphi = dm - d$ with some constant d. The theorem is proved. $\qquad\square$

11.4   Superstability of generalized moment functions on hypergroups

Let K be a hypergroup. In this section we prove that generalized moment functions possess a remarkable superstability property.

Theorem 11.7. *Let* K *be hypergroup,* n *a nonnegative integer,* $\epsilon > 0$ *a real number and suppose that for the unbounded functions* $f_k : K \to \mathbb{C}$ *$(k = 0, 1, \ldots, n)$ the functions*

$$(x, y) \mapsto f_k(x * y) - \sum_{j=0}^{k} \binom{k}{j} f_j(x) f_{k-j}(y)$$

are bounded on $K \times K$. Then the sequence $(f_k)_{k \leq n}$ forms a moment function sequence of order n on K.

Proof. We prove the theorem for a fixed n by induction on k. For $k = 0$ we have by assumption that the function

$$(x, y) \mapsto f_0(x * y) - f_0(x) f_0(y)$$

is bounded on $K \times K$. By Theorem 11.1 this implies that f_0 is an exponential on K. Suppose now that $k \geq 1$ and we have proved that the functions f_j for $j = 0, 1, \ldots, k - 1$ form a moment function sequence of order $k - 1$ on K. By assumption we have that the function

$$(x, y, z) \mapsto F(x, y, z) = f_k(x * y * z) - \sum_{j=0}^{k} \binom{k}{j} f_j(x * y) f_{k-j}(z)$$

and also the function

$$(x, y, z) \mapsto G(x, y, z) = f_k(x * y * z) - \sum_{j=0}^{k} \binom{k}{j} f_j(x) f_{k-j}(y * z)$$

are bounded on $K \times K \times K$. Then their difference

$$(x, y, z) \mapsto F(x, y, z) - G(x, y, z)$$

$$= \sum_{j=0}^{k} \binom{k}{j} f_j(x * y) f_{k-j}(z) - \sum_{j=0}^{k} \binom{k}{j} f_j(x) f_{k-j}(y * z)$$

is also bounded. By our induction hypothesis this means that the function

$$(x, y, z) \mapsto F(x, y, z) - G(x, y, z) = H(x, y, z)$$

$$= \sum_{j=1}^{k-1} \binom{k}{j} f_j(x) \sum_{i=0}^{k-j} \binom{k-j}{i} f_i(y) f_{k-j-i}(z)$$

$$- \sum_{j=1}^{k-1} \binom{k}{j} \sum_{i=0}^{j} \binom{j}{i} f_i(x) f_{j-i}(y) f_{k-j}(z)$$

$$+ f_0(x) f_k(y * z) - f_k(x * y) f_0(z) + f_k(x) f_0(y) f_0(z) - f_0(x) f_0(y) f_k(z)$$

is bounded, too. We can reorder the terms of this sum in the following way:

$$H(x, y, z) = f_0(x) \Big[f_k(y * z) - \sum_{j=0}^{k} \binom{k}{j} f_j(y) f_{k-j}(z) \Big]$$

$$- f_0(z) \Big[f_k(x * y) - \sum_{j=0}^{k} \binom{k}{j} f_j(x) f_{k-j}(y) \Big]$$

$$+ \sum_{j=1}^{k-1} \sum_{i=0}^{k-j-1} \binom{k}{j} \binom{k-j}{i} f_j(x) f_i(y) f_{k-j-i}(z)$$

$$- \sum_{j=1}^{k-1} \sum_{i=1}^{j} \binom{k}{j} \binom{j}{i} f_i(x) f_{j-i}(y) f_{k-j}(z)$$

for all x, y, z in K. We show that the two terms on the right hand side of the last equality cancel. In the first term replacing i by $t - j$ and in the second term interchanging the sums we obtain

$$H(x, y, z) = f_0(x) \Big[f_k(y * z) - \sum_{j=0}^{k} \binom{k}{j} f_j(y) f_{k-j}(z) \Big]$$

$$- f_0(z) \Big[f_k(x * y) - \sum_{j=0}^{k} \binom{k}{j} f_j(x) f_{k-j}(y) \Big]$$

$$+ \sum_{j=1}^{k-1} \sum_{t=j}^{k-1} \binom{k}{j} \binom{k-j}{k-t} f_j(x) f_{t-j}(y) f_{k-t}(z)$$

$$-\sum_{i=1}^{k-1}\sum_{j=i}^{k-1}\binom{k}{j}\binom{j}{i}f_i(x)f_{j-i}(y)f_{k-j}(z)$$

for all x, y, z in K. In the second term we write j for i and t for j to get

$$H(x, y, z) = f_0(x)\Big[f_k(y * z) - \sum_{j=0}^{k}\binom{k}{j}f_j(y)f_{k-j}(z)\Big]$$

$$-f_0(z)\Big[f_k(x * y) - \sum_{j=0}^{k}\binom{k}{j}f_j(x)f_{k-j}(y)\Big]$$

$$+\sum_{j=1}^{k-1}\sum_{t=j}^{k-1}\binom{k}{j}\binom{k-j}{k-t}f_j(x)f_{t-j}(y)f_{k-t}(z)$$

$$-\sum_{j=1}^{k-1}\sum_{t=j}^{k-1}\binom{k}{t}\binom{t}{j}f_j(x)f_{t-j}(y)f_{k-t}(z)$$

for all x, y, z in K. On the other hand, we have

$$\binom{k}{j}\binom{k-j}{k-t} = \frac{k!}{j!(k-j)!}\frac{(k-j)!}{(k-t)!(t-j)!} = \frac{k!}{t!(k-t)!}\frac{t!}{j!(t-j)!} = \binom{k}{t}\binom{t}{j},$$

hence the function

$$(x, y, z) \mapsto L(x, y, z) = f_0(x)\Big[f_k(y * z) - \sum_{j=0}^{k}\binom{k}{j}f_j(y)f_{k-j}(z)\Big]$$

$$-f_0(z)\Big[f_k(x * y) - \sum_{j=0}^{k}\binom{k}{j}f_j(x)f_{k-j}(y)\Big]$$

is bounded. If there are y, z in K such that

$$f_k(y * z) - \sum_{j=0}^{k} \binom{k}{j} f_j(y) f_{k-j}(z) \neq 0 \,,$$

then f_0 is bounded, which is impossible. Thus we have

$$f_k(y * z) - \sum_{j=0}^{k} \binom{k}{j} f_j(y) f_{k-j}(z) = 0$$

for all y, z in K and the proof is complete. $\square$

Bibliography

M. Amini and C. H. Chu. Harmonic functions on hypergroups. *J. Funct. Anal.*, 261(7):1835–1864, 2011.

J. Aczél. *Lectures on functional equations and their applications.* Mathematics in Science and Engineering, Vol. 19. Academic Press, New York, 1966. Translated by Scripta Technica, Inc. Supplemented by the author. Edited by Hansjorg Oser.

J Aczél. Functions of binomial type mapping groupoids into rings. *Math. Z.*, 154(2):115–124, 1977.

J. Aczél and J. Dhombres. *Functional equations in several variables*, volume 31 of *Encyclopedia of Mathematics and its Applications.* Cambridge University Press, Cambridge, 1989. With applications to mathematics, information theory and to the natural and social sciences.

H. A. Ali and F. M. Kandil. Convergence of exponential convex functions on hypergroups. *Far East J. Math. Sci. (FJMS)*, 43(2):225–234, 2010.

N. I. Akhiezer. *The classical moment problem and some related questions in analysis.* Translated by N. Kemmer. Hafner Publishing Co., New York, 1965.

M. Amini. Fourier transform of unbounded measures on hypergroups. *Boll. Unione Mat. Ital. Sez. B Artic. Ric. Mat. (8)*, 10(3, bis):819–828, 2007.

J. An and D. Yang. A Levi-Civitá equation on compact groups and nonabelian Fourier analysis. *Integral Equations Operator Theory*, 66(2):183–195, 2010.

A. Azimifard. On multipliers for the Hilbert space of a hypergroup. *C. R. Math. Acad. Sci. Soc. R. Can.*, 30(3):84–88, 2008.

A. Azimifard. α-amenable hypergroups. *Math. Z.*, 265(4):971–982, 2010.

J. J. Betancor, J. D. Betancor, and J. M. R. Méndez. Chébli-Trimèche hypergroups and W-type spaces. *J. Math. Anal. Appl.*, 271(2):359–373, 2002.

J. J. Benedetto. *Spectral synthesis.* Academic Press Inc. [Harcourt Brace Jovanovich Publishers], New York, 1975. Pure and Applied Mathematics, No. 66.

A. Beurling. On the spectral synthesis of bounded functions. *Acta Math.*, 81:14, 1948.

W. R. Bloom and H. Heyer. *Harmonic analysis of probability measures on hyper-*

groups, volume 20 of *de Gruyter Studies in Mathematics*. Walter de Gruyter & Co., Berlin, 1995.

W. R. Bloom and H. Heyer. Negative definite functions and convolution semigroups of probability measures on a commutative hypergroup. *Probab. Math. Statist.*, 16(1):157–176, 1996.

W. R. Bloom and H. Heyer. Polynomial hypergroup structures and applications to probability theory. *Publ. Math. Debrecen*, 72(1-2):199–225, 2008.

J. E. Björk. *Rings of differential operators*, volume 21 of *North-Holland Mathematical Library*. North-Holland Publishing Co., Amsterdam, 1979.

Yu. M. Berezansky and A. A. Kalyuzhnyĭ. Hypercomplex systems and hypergroups: connections and distinctions. In *Applications of hypergroups and related measure algebras (Seattle, WA, 1993)*, volume 183 of *Contemp. Math.*, pages 21–44. Amer. Math. Soc., Providence, RI, 1995.

W. R. Bloom and P. Ressel. Exponentially bounded positive-definite functions on a commutative hypergroup. *J. Austral. Math. Soc. Ser. A*, 61(2):238–248, 1996.

W. R. Bloom and P. Ressel. Representations of negative definite functions on polynomial hypergroups. *Arch. Math. (Basel)*, 78(4):318–328, 2002.

W. R. Bloom and P. Ressel. Negative definite and Schoenberg functions on commutative hypergroups. *J. Aust. Math. Soc.*, 79(1):25–37, 2005.

W. R. Bloom and Z. Xu. The Hardy-Littlewood maximal function for Chébli-Trimèche hypergroups. In *Applications of hypergroups and related measure algebras (Seattle, WA, 1993)*, volume 183 of *Contemp. Math.*, pages 45–70. Amer. Math. Soc., Providence, RI, 1995.

W. R. Bloom and Z. Xu. Hardy spaces on Chébli-Trimèche hypergroups. *Methods Funct. Anal. Topology*, 3(2):1–26, 1997.

W. R. Bloom and Z. Xu. Fourier multipliers for local Hardy spaces on Chébli-Trimèche hypergroups. *Canad. J. Math.*, 50(5):897–928, 1998.

W. R. Bloom and Z. Xu. Fourier transforms of Schwartz functions on Chébli-Trimèche hypergroups. *Monatsh. Math.*, 125(2):89–109, 1998.

W. R. Bloom and Z. Xu. Fourier multipliers for L^p on Chébli-Trimèche hypergroups. *Proc. London Math. Soc. (3)*, 80(3):643–664, 2000.

W. R. Bloom and Z. Xu. Maximal functions on Chébli-Trimèche hypergroups. *Infin. Dimens. Anal. Quantum Probab. Relat. Top.*, 3(3):403–434, 2000.

William C. Connett, Marc-Olivier Gebuhrer, and Alan L. Schwartz, editors. *Applications of hypergroups and related measure algebras*, volume 183 of *Contemporary Mathematics*, Providence, RI, 1995. American Mathematical Society.

H. Chébli. Sturm-Liouville hypergroups. In *Applications of hypergroups and related measure algebras (Seattle, WA, 1993)*, volume 183 of *Contemp. Math.*, pages 71–88. Amer. Math. Soc., Providence, RI, 1995.

A. K. Chilana and A. Kumar. Spectral synthesis in Segal algebras on hypergroups. *Pacific J. Math.*, 80(1):59–76, 1979.

J. K. Chung, Pl. Kannappan, and C. T. Ng. A generalization of the cosine-sine functional equation on groups. *Linear Algebra Appl.*, 66:259–277, 1985.

J. B. Conway. *A course in functional analysis*, volume 96 of *Graduate Texts in*

Mathematics. Springer-Verlag, New York, second edition, 1990.

I. Corovei. The cosine functional equation for nilpotent groups. *Aequationes Math.*, 15(1):99–106, 1977.

C. Corduneanu. *Functional equations with causal operators*, volume 16 of *Stability and Control: Theory, Methods and Applications.* Taylor & Francis, London, 2002.

A. K. Chilana and K. A. Ross. Spectral synthesis in hypergroups. *Pacific J. Math.*, 76(2):313–328, 1978.

E. Castillo and M. R. Ruiz-Cobo. *Functional equations and modelling in science and engineering*, volume 161 of *Monographs and Textbooks in Pure and Applied Mathematics.* Marcel Dekker Inc., New York, 1992.

W. C. Connett and A. L. Schwartz. Continuous 2-variable polynomial hypergroups. In *Applications of hypergroups and related measure algebras (Seattle, WA, 1993)*, volume 183 of *Contemp. Math.*, pages 89–109. Amer. Math. Soc., Providence, RI, 1995.

W. C. Connett and A. L. Schwartz. Subsets of **R** which support hypergroups with polynomial characters. In *Proceedings of the International Conference on Orthogonality, Moment Problems and Continued Fractions (Delft, 1994)*, volume 65, pages 73–84, 1995.

W. C. Connett and A. L. Schwartz. Hypergroups and differential equations. In *Lie groups and Lie algebras*, volume 433 of *Math. Appl.*, pages 109–115. Kluwer Acad. Publ., Dordrecht, 1998.

Yu. A. Chapovsky and L. I. Vainerman. Compact quantum hypergroups. *J. Operator Theory*, 41(2):261–289, 1999.

S. Czerwik. *Functional equations and inequalities in several variables.* World Scientific Publishing Co. Inc., River Edge, NJ, 2002.

T.M.K. Davison. D'Alembert's functional equation and Chebyshev polynomials. *Ann. Acad. Paed. Cracov., Studia Math.*, 4:31–38, 2001.

V. Ditkin. On the structure of ideals in certain normed rings. *Uchenye Zapiski Moskov. Gos. Univ. Matematika*, 30:83–130, 1939.

R. Daher and T. Kawazoe. An uncertainty principle on Sturm-Liouville hypergroups. *Proc. Japan Acad. Ser. A Math. Sci.*, 83(9-10):167–169, 2007.

N. Dunford and J. T. Schwartz. *Linear operators. Part I.* Wiley Classics Library. John Wiley & Sons Inc., New York, 1988. General theory, With the assistance of William G. Bade and Robert G. Bartle, Reprint of the 1958 original, A Wiley-Interscience Publication.

N. Dunford and J. T. Schwartz. *Linear operators. Part II.* Wiley Classics Library. John Wiley & Sons Inc., New York, 1988. Spectral theory. Selfadjoint operators in Hilbert space, With the assistance of William G. Bade and Robert G. Bartle, Reprint of the 1963 original, A Wiley-Interscience Publication.

N. Dunford and J. T. Schwartz. *Linear operators. Part III.* Wiley Classics Library. John Wiley & Sons Inc., New York, 1988. Spectral operators, With the assistance of William G. Bade and Robert G. Bartle, Reprint of the 1971 original, A Wiley-Interscience Publication.

L. Ehrenpreis. Appendix to the paper "Mean periodic functions I". *Amer. J. Math.*, 77:731–733, 1955.

L. Ehrenpreis. Mean periodic functions. I. Varieties whose annihilator ideals are principal. *Amer. J. Math.*, 77:293–328, 1955.

M. Ehring. A large deviation principle for polynomial hypergroups. *J. London Math. Soc. (2)*, 53(1):197–208, 1996.

K. Ey and R. Lasser. Facing linear difference equations through hypergroup methods. *J. Difference Equ. Appl.*, 13(10):953–965, 2007.

R. J. Elliott. Two notes on spectral synthesis for discrete Abelian groups. *Proc. Cambridge Philos. Soc.*, 61:617–620, 1965.

M. Engert. Finite dimensional translation invariant subspaces. *Pacific J. Math.*, 32:333–343, 1970.

J. Favard. Remarque sur les polynomes trigonométriques. *C. R. Acad. Sci. Paris*, 209:746–748, 1939.

G. Feldman. *Functional equations and characterization problems on locally compact abelian groups*, volume 5 of *EMS Tracts in Mathematics*. European Mathematical Society (EMS), Zürich, 2008.

Y. Funakoshi and S. Kawakami. Entropy of probability measures on finite commutative hypergroups. *Bull. Nara Univ. Ed. Natur. Sci.*, 57(2):17–20, 2008.

F. Filbir and R. Lasser. Reiter's condition P_2 and the Plancherel measure for hypergroups. *Illinois J. Math.*, 44(1):20–32, 2000.

P. G. A. Floris. A noncommutative discrete hypergroup associated with q-disk polynomials. *J. Comput. Appl. Math.*, 68(1-2):69–78, 1996.

F. Filbir, R. Lasser, and R. Szwarc. Hypergroups of compact type. *J. Comput. Appl. Math.*, 178(1-2):205–214, 2005.

B. Forte. *Functional Equations and Inequalities*. Springer, Berlin, Heidelberg, 2010.

W. Förg-Rob and J. Schwaiger. A generalization of the cosine equation to n summands. In *Selected topics in functional equations and iteration theory (Graz, 1991)*, volume 316 of *Grazer Math. Ber.*, pages 219–226. Karl-Franzens-Univ. Graz, Graz, 1992.

L. Gallardo. Asymptotic drift of the convolution and moment functions on hypergroups. *Math. Z.*, 224(3):427–444, 1997.

L. Gallardo. Some methods to find moment functions on hypergroups. In *Harmonic analysis and hypergroups (Delhi, 1995)*, Trends Math., pages 13–31. Birkhäuser Boston, Boston, MA, 1998.

L. Gallardo. A central limit theorem for Markov chains and applications to hypergroups. *Proc. Amer. Math. Soc.*, 127(6):1837–1845, 1999.

L. Gallardo. A multidimensional central limit theorem for random walks on hypergroups. *Stoch. Stoch. Rep.*, 73(1-2):1–23, 2002.

M. O. Gebuhrer. Haar measure on locally compact hypergroups. In *Lie groups and Lie algebras*, volume 433 of *Math. Appl.*, pages 117–131. Kluwer Acad. Publ., Dordrecht, 1998.

A. Ghaffari. Weakly almost periodic functions on hypergroup algebras. *Far East J. Math. Sci. (FJMS)*, 5(3):277–287, 2002.

J. E. Gilbert. Spectral synthesis problems for invariant subspaces on groups. I. *Amer. J. Math.*, 88:626–635, 1966.

F. P. Greenleaf. *Invariant means on topological groups and their applications.*

Van Nostrand Mathematical Studies, No. 16. Van Nostrand Reinhold Co., New York, 1969.

M. O. Gebuhrer and R. Szwarc. On symmetry of discrete polynomial hypergroups. *Proc. Amer. Math. Soc.*, 127(6):1705–1709, 1999.

M.O. Gebuhrer and A. L. Schwartz. Harmonic analysis on compact commutative hypergroups: the role of the maximum subgroup. *J. Anal. Math.*, 82:175–206, 2000.

L. Gallardo and K. Trimèche. Lie theorems for one-dimensional hypergroups. *Integral Transforms Spec. Funct.*, 13(1):71–92, 2002.

L. Gallardo and K. Trimèche. One dimensional diffusive hypergroups with asymptotic drift. *Integral Transforms Spec. Funct.*, 13(2):101–108, 2002.

L. B. Hanes. *Walsh-Fourier series on the hypergroup deformation of the dyadic group*. ProQuest LLC, Ann Arbor, MI, 1994. Thesis (Ph.D.)–University of Oregon.

F. Hausdorff. Summationsmethoden und Momentfolgen. I. *Math. Z.*, 9(1-2):74–109, 1921.

F. Hausdorff. Summationsmethoden und Momentfolgen. II. *Math. Z.*, 9(3-4):280–299, 1921.

H. Helson. Spectral synthesis of bounded functions. *Ark. Mat.*, 1:497–502, 1952.

H. Helson. *Harmonic analysis*. Addison-Wesley Publishing Company Advanced Book Program, Reading, MA, 1983.

P. Hermann. Representations of double coset hypergroups and induced representations. *Manuscripta Math.*, 88(1):1–24, 1995.

H. Heyer. Functional limit theorems for random walks on one-dimensional hypergroups. In *Stability problems for stochastic models (Suzdal, 1991)*, volume 1546 of *Lecture Notes in Math.*, pages 45–57. Springer, Berlin, 1993.

H. Heyer. Progress in the theory of probability on hypergroups. In *Applications of hypergroups and related measure algebra (Seattle, WA, 1993)*, volume 183 of *Contemp. Math.*, pages 191–212. Amer. Math. Soc., Providence, RI, 1995.

H. Heyer. Positive and negative definite functions on a hypergroup and its dual. In *Infinite dimensional harmonic analysis IV*, pages 63–96. World Sci. Publ., Hackensack, NJ, 2009.

V. Hösel, M. Hofmann, and R. Lasser. Means and Følner condition on polynomial hypergroups. *Mediterr. J. Math.*, 7(1):75–88, 2010.

J. Hinz. Hypergroup actions and wavelets. In *Infinite dimensional harmonic analysis (Kyoto, 1999)*, pages 167–176. Gräbner, Altendorf, 2000.

D. H. Hyers, G. Isac, and Th. M. Rassias. *Stability of functional equations in several variables*. Progress in Nonlinear Differential Equations and their Applications, 34. Birkhäuser Boston Inc., Boston, MA, 1998.

H. Heyer and S. Kawakami. Extensions of Pontryagin hypergroups. *Probab. Math. Statist.*, 26(2):245–260, 2006.

V. Hösel and R. Lasser. Prediction of weakly stationary sequences on polynomial hypergroups. *Ann. Probab.*, 31(1):93–114, 2003.

J. Huang and H. Liu. An analogue of Beurling's theorem for the Laguerre hypergroup. *J. Math. Anal. Appl.*, 336(2):1406–1413, 2007.

M. Hosszú. A remark on the square norm. *Aequationes Math.*, 2:190–193, 1969.

H. Heyer and Gy. Pap. Martingale characterizations of increment processes in a commutative hypergroup. *Adv. Pure Appl. Math.*, 1(1):117–140, 2010.

E. Hewitt and K. A. Ross. *Abstract harmonic analysis. Vol. I: Structure of topological groups. Integration theory, group representations*. Die Grundlehren der mathematischen Wissenschaften, Bd. 115. Academic Press Inc., Publishers, New York, 1963.

E. Hewitt and K. A. Ross. *Abstract harmonic analysis. Vol. I*, volume 115 of *Grundlehren der Mathematischen Wissenschaften [Fundamental Principles of Mathematical Sciences]*. Springer-Verlag, Berlin, second edition, 1979. Structure of topological groups, integration theory, group representations.

P. Hermann and M. Voit. Induced representations and duality results for commutative hypergroups. *Forum Math.*, 7(5):543–558, 1995.

A. J. Izzo. A functional analysis proof of the existence of Haar measure on locally compact abelian groups. *Proc. Amer. Math. Soc.*, 115(2):581–583, 1992.

A. Járai. On regular solutions of functional equations. *Aequationes Math.*, 30(1):21–54, 1986.

A. Járai. *Regularity properties of functional equations in several variables*, volume 8 of *Advances in Mathematics (Springer)*. Springer, New York, 2005.

R. I. Jewett. Spaces with an abstract convolution of measures. *Advances in Math.*, 18(1):1–101, 1975.

A. Járai and L. Székelyhidi. Regularization and general methods in the theory of functional equations. *Aequationes Math.*, 52(1-2):10–29, 1996.

L. Jaafar and K. Trimèche. Wavelets on the product of Euclidean hypergroup and Chébli-Trimèche hypergroup. In *Functional analysis*, pages 203–220. Narosa, New Delhi, 1998.

Pl. Kannappan. *Functional equations and inequalities with applications*. Springer Monographs in Mathematics. Springer, New York, 2009.

I. Kaplansky. Primary ideals in group algebras. *Proc. Nat. Acad. Sci. U. S. A.*, 35:133–136, 1949.

R. A. Kamyabi-Gol. P-amenable locally compact hypergroups. *Bull. Iranian Math. Soc.*, 32(2):43–51, 2006.

A. A. Kalyuzhnyi, G. B. Podkolzin, and Yu. A. Chapovsky. Harmonic analysis on a locally compact hypergroup. *Methods Funct. Anal. Topology*, 16(4):304–332, 2010.

A. Kumar and A. I. Singh. Spectral synthesis in products and quotients of hypergroups. *Pacific J. Math.*, 94(1):177–192, 1981.

T. H. Koornwinder and A. L. Schwartz. Product formulas and associated hypergroups for orthogonal polynomials on the simplex and on a parabolic biangle. *Constr. Approx.*, 13(4):537–567, 1997.

M. Kuczma. *An introduction to the theory of functional equations and inequalities*. Birkhäuser Verlag, Basel, second edition, 2009. Cauchy's equation and Jensen's inequality, Edited and with a preface by Attila Gilányi.

A. Kumar. A qualitative uncertainty principle for certain hypergroups. *Glas. Mat. Ser. III*, 36(56)(1):33–38, 2001.

P. G. Laird. Some properties of mean periodic functions. *J. Austral. Math. Soc.*,

14:424–432, 1972.

R. Lasser. Orthogonal polynomials and hypergroups. *Rend. Mat. (7)*, 3(2):185–209, 1983.

R. Lasser. Orthogonal polynomials and hypergroups. II. The symmetric case. *Trans. Amer. Math. Soc.*, 341(2):749–770, 1994.

R. Lasser. On the character space of commutative hypergroups. *Jahresber. Deutsch. Math.-Verein.*, 104(1):3–16, 2002.

R. Lasser. Discrete commutative hypergroups. In *Inzell Lectures on Orthogonal Polynomials*, volume 2 of *Adv. Theory Spec. Funct. Orthogonal Polynomials*, pages 55–102. Nova Sci. Publ., Hauppauge, NY, 2005.

R. Lasser. Amenability and weak amenability of l^1-algebras of polynomial hypergroups. *Studia Math.*, 182(2):183–196, 2007.

R. Lasser. Point derivations on the L^1-algebra of polynomial hypergroups. *Colloq. Math.*, 116(1):15–30, 2009.

R. Lasser. Various amenability properties of the L^1-algebra of polynomial hypergroups and applications. *J. Comput. Appl. Math.*, 233(3):786–792, 2009.

A. T. M. Lau. Analysis on a class of Banach algebras with applications to harmonic analysis on locally compact groups and semigroups. *Fund. Math.*, 118(3):161–175, 1983.

M. Lashkarizadeh Bami, M. Pourgholamhossein, and H. Samea. Fourier algebras on locally compact hypergroups. *Math. Nachr.*, 282(1):16–25, 2009.

M. Lashkarizadeh Bami and H. Samea. Amenability and essential amenability of certain convolution Banach algebras on compact hypergroups. *Bull. Belg. Math. Soc. Simon Stevin*, 16(1):145–152, 2009.

M. Lefranc. Analyse spectrale sur $\mathbf{Z_n}$. *C. R. Acad. Sci. Paris*, 246:1951–1953, 1958.

L. H. Loomis. *An introduction to abstract harmonic analysis*. D. Van Nostrand Company, Inc., Toronto-New York-London, 1953.

R. Lasser, J. Obermaier, and H. Rauhut. Generalized hypergroups and orthogonal polynomials. *J. Aust. Math. Soc.*, 82(3):369–393, 2007.

L. Losonczi. An extension theorem for the Levi-Civita functional equation and its applications. In *Contributions to the theory of functional equations (Graz, 1991)*, volume 315 of *Grazer Math. Ber.*, pages 51–68. Karl-Franzens-Univ. Graz, Graz, 1991.

R. Lasser and M. Skantharajah. Reiter's condition for amenable hypergroups. *Monatsh. Math.*, 163(3):327–338, 2011.

M. N. Lazhari and K. Trimèche. Convolution algebras and factorization of measures on Chébli-Trimèche hypergroups. *C. R. Math. Rep. Acad. Sci. Canada*, 17(4):165–169, 1995.

S. Mandelbrojt and S. Agmon. Une généralisation du théorème tauberien de Wiener. *Acta Sci. Math. Szeged*, 12(Leopoldo Fejer et Frederico Riesz LXX annos natis dedicata, Pars B):167–176, 1950.

R. Ma. Heisenberg uncertainty principle on Chébli-Trimèche hypergroups. *Pacific J. Math.*, 235(2):289–296, 2008.

B. Malgrange. Sur quelques propriétés des équations de convolution. *C. R. Acad. Sci. Paris*, 238:2219–2221, 1954.

P. Malliavin. Impossibilité de la synthèse spectrale sur les groupes abéliens non compacts. *Inst. Hautes Études Sci. Publ. Math.*, 1959:85–92, 1959.

M. A. McKiernan. Measurable solutions of quadratic functional equations. *Colloq. Math.*, 35(1):97–103, 1976.

M. A. McKiernan. Equations of the form $H(x \circ y) = \sum_i f_i(x)g_i(y)$. *Aequationes Math.*, 16(1-2):51–58, 1977.

M. A. McKiernan. The matrix equation $a(x \circ y) = a(x) + a(x)a(y) + a(y)$. *Aequationes Math.*, 15(2-3):213–223, 1977.

S. Menges. Functional limit theorems for probability measures on hypergroups. *Probab. Math. Statist.*, 25(1, Acta Univ. Wratislav. No. 2784):155–171, 2005.

W. Młotkowski. Some class of polynomial hypergroups. In *Quantum probability*, volume 73 of *Banach Center Publ.*, pages 357–362. Polish Acad. Sci. Inst. Math., Warsaw, 2006.

V. Muruganandam. Fourier algebra of a hypergroup. I. *J. Aust. Math. Soc.*, 82(1):59–83, 2007.

V. Muruganandam. Fourier algebra of a hypergroup. II. Spherical hypergroups. *Math. Nachr.*, 281(11):1590–1603, 2008.

A. Nasr-Isfahani. Representations and positive definite functions on hypergroups. *Serdica Math. J.*, 25(4):283–296, 1999.

A. Nasr-Isfahani. Bochner theorems for commutative locally compact hypergroups. In *Proceedings of the 31st Iranian Mathematics Conference (Tehran, 2000)*, pages 271–274. Univ. Tehran, Tehran, 2000.

R. Nasr-Isfahani. On exponentially bounded positive-definite functions on hypergroups. *Arch. Math. (Basel)*, 76(6):455–457, 2001.

R. Nasr-Isfahani. Integral representation for exponentially bounded negative-definite functions on hypergroups. *Math. Nachr.*, 256:82–87, 2003.

M. M. Nessibi, L. T. Rachdi, and K. Trimèche. The local central limit theorem on the product of the Chébli-Trimèche hypergroup and the Euclidean hypergroup $\mathbf{R}^n$. *J. Math. Sci. (Calcutta)*, 9(2):109–123, 1998.

M. M. Nessibi and M. Sifi. Laguerre hypergroup and limit theorem. In *Lie groups and Lie algebras*, volume 433 of *Math. Appl.*, pages 133–145. Kluwer Acad. Publ., Dordrecht, 1998.

M. M. Nessibi and B. Selmi. A Wiener-Tauberian and a Pompeiu type theorems on the Laguerre hypergroup. *J. Math. Anal. Appl.*, 351(1):232–243, 2009.

A. S. Okb El-Bab and F. M. Bayumi. Spectrum of positive definite functions on hypergroups. *Kyungpook Math. J.*, 35(1):17–23, 1995.

A. S. Okb El Bab and H. A. Ghany. Harmonic analysis on hypergroups. *Int. J. Pure Appl. Math.*, 64(1):9–19, 2010.

P. Oleszczuk. Laguerre entire functions and Sturm-Liouville hypergroups. *Methods Funct. Anal. Topology*, 7(3):67–79, 2001.

Á. Orosz. Difference equations on discrete polynomial hypergroups. *Adv. Difference Equ.*, pages Art. 51427, 1–10, 2006.

Á. Orosz. Sine and cosine equation on discrete polynomial hypergroups. *Aequationes Math.*, 72(3):225–233, 2006.

Á. Orosz and L. Székelyhidi. Moment functions on polynomial hypergroups in several variables. *Publ. Math. Debrecen*, 65(3-4):429–438, 2004.

Á. Orosz and L. Székelyhidi. Moment functions on polynomial hypergroups. *Arch. Math. (Basel)*, 85(2):141–150, 2005.

Á. Orosz and L. Székelyhidi. Moment functions on Sturm-Liouville hypergroups. *Ann. Univ. Sci. Budapest. Sect. Comput.*, 29:141–156, 2008.

N. Obata and N. J. Wildberger. Generalized hypergroups and orthogonal polynomials. *Nagoya Math. J.*, 142:67–93, 1996.

A. W. Parr. *Signed hypergroups*. ProQuest LLC, Ann Arbor, MI, 1997. Thesis (Ph.D.)–University of Oregon.

L. Pavel. On quasi-positive definite functions and representations of hypergroups in QP_n spaces. *Rocky Mountain J. Math.*, 27(3):889–902, 1997.

L. Pavel. Multipliers and induced multiplier representations on hypergroups. *Math. Rep. (Bucur.)*, 1(51)(4):601–609, 1999.

L. Pavel. Positive definite measures on hypergroups. *An. Univ. Bucureşti Mat.*, 48(1):51–56, 1999.

L. Pavel. Ergodic sequences of probability measures on commutative hypergroups. *Int. J. Math. Math. Sci.*, (5-8):335–343, 2004.

L. Pavel. An ergodic property of amenable hypergroups. *Bol. Soc. Mat. Mexicana (3)*, 13(1):123–129, 2007.

L. Pavel. On trigonometric polynomials in the L_1-algebra of compact hypergroups. *An. Univ. Bucureşti Mat.*, 56(2):253–260, 2007.

L. Pavel. Weak containment and hypergroup algebras. *Rev. Roumaine Math. Pures Appl.*, 52(1):87–93, 2007.

C. R. E. Raja. Normed convergence property for hypergroups admitting an invariant measure. *Southeast Asian Bull. Math.*, 26(3):479–481, 2002.

K. A. Ross, J. M. Anderson, G. L. Litvinov, A. I. Singh, V. S. Sunder, and Martin E. Walter, editors. *Harmonic analysis and hypergroups*, Trends in Mathematics, Boston, MA, 1998. Birkhäuser Boston Inc.

C. Rentzsch. Canonical representation of convolution semigroups on hypergroups. In *Infinite-dimensional harmonic analysis (Tübingen, 1995)*, pages 188–194. Gräbner, Tübingen, 1996.

C. Rentzsch. A Lévy Khintchine type representation of convolution semigroups on commutative hypergroups. *Probab. Math. Statist.*, 18(1, Acta Univ. Wratislav. No. 2045):185–198, 1998.

J. Riss. Transformation de Fourier des distributions. *C. R. Acad. Sci. Paris*, 229:12–14, 1949.

K. A. Ross. Hypergroups and centers of measure algebras. In *Symposia Mathematica, Vol. XXII (Convegno sull'Analisi Armonica e Spazi di Funzioni su Gruppi Localmente Compatti, INDAM, Rome, 1976)*, pages 189–203. Academic Press, London, 1977.

M. Rösler. Bessel-type signed hypergroups on **R**. In *Probability measures on groups and related structures, XI (Oberwolfach, 1994)*, pages 292–304. World Sci. Publ., River Edge, NJ, 1995.

M. Rösler. On the dual of a commutative signed hypergroup. *Manuscripta Math.*, 88(2):147–163, 1995.

K. A. Ross. Signed hypergroups—a survey. In *Applications of hypergroups and re-*

lated measure algebras (Seattle, WA, 1993), volume 183 of *Contemp. Math.*, pages 319–329. Amer. Math. Soc., Providence, RI, 1995.

K. A. Ross. Hypergroups and signed hypergroups. In *Harmonic analysis and hypergroups (Delhi, 1995)*, Trends Math., pages 77–91. Birkhäuser Boston, Boston, MA, 1998.

K. A. Ross. LCA hypergroups. In *Proceedings of the 14th Summer Conference on General Topology and its Applications (Brookville, NY, 1999)*, volume 24, pages 533–546 (2001), 1999.

R. A. Rosenbaum and S. L. Segal. A functional equation characterising the sine. *Math. Gaz.*, 44:97–105, 1960.

W. Rudin. *Functional analysis.* McGraw-Hill Book Co., New York, 1973. McGraw-Hill Series in Higher Mathematics.

M. Rösler and M. Voit. Partial characters and signed quotient hypergroups. *Canad. J. Math.*, 51(1):96–116, 1999.

K. A. Ross and D. Xu. Norm convergence of random walks on compact hypergroups. *Math. Z.*, 214(3):415–423, 1993.

K. A. Ross and D. Xu. Hypergroup deformations and Markov chains. *J. Theoret. Probab.*, 7(4):813–830, 1994.

A. Roukbi and D. Zeglami. d'Alembert's functional equations on hypergroups. *Adv. Pure Appl. Math.*, 2(2):147–166, 2011.

H. Samea. Weak amenability of convolution Banach algebras on compact hypergroups. *Bull. Korean Math. Soc.*, 47(2):307–317, 2010.

L. Schwartz. Théorie générale des fonctions moyenne-périodiques. *Ann. of Math. (2)*, 48:857–929, 1947.

L. Schwartz. Sur une propriété de synthèse spectrale dans les groupes non compacts. *C. R. Acad. Sci. Paris*, 227:424–426, 1948.

A. L. Schwartz. Three lectures on hypergroups: Delhi, December 1995. In *Harmonic analysis and hypergroups (Delhi, 1995)*, Trends Math., pages 93–129. Birkhäuser Boston, Boston, MA, 1998.

S. L. Segal. On a sine functional equation. *Amer. Math. Monthly*, 70:306–308, 1963.

J. Shohat. The Relation of the Classical Orthogonal Polynomials to the Polynomials of Appell. *Amer. J. Math.*, 58(3):453–464, 1936.

E. Shulman. Some extensions of the Levi-Civitá functional equation and richly periodic spaces of functions. *Aequationes Math.*, 81(1-2):109–120, 2011.

A. I. Singh. Completely positive hypergroup actions. *Mem. Amer. Math. Soc.*, 124(593):xii+68, 1996.

P. K. Sahoo and P. Kannappan. *Introduction to functional equations.* CRC Press, Boca Raton, FL, 2011.

P. K. Sahoo and T. Riedel. *Mean value theorems and functional equations.* World Scientific Publishing Co. Inc., River Edge, NJ, 1998.

H. Stetkaer. On a signed cosine equation of N summands. *Aequationes Math.*, 51(3):294–302, 1996.

T. J. Stieltjes. Recherches sur les fractions continues. *Ann. Fac. Sci. Toulouse Sci. Math. Sci. Phys.*, 8(4):J1–J122, 1894.

B. Sturmfels. *Solving systems of polynomial equations*, volume 97 of *CBMS Re-*

gional Conference Series in Mathematics. Published for the Conference Board of the Mathematical Sciences, Washington, DC, 2002.

V. S. Sunder and N. J. Wildberger. Actions of finite hypergroups and examples. In *Harmonic analysis and hypergroups (Delhi, 1995)*, Trends Math., pages 145–163. Birkhäuser Boston, Boston, MA, 1998.

V. S. Sunder and N. J. Wildberger. Actions of finite hypergroups. *J. Algebraic Combin.*, 18(2):135–151, 2003.

L. Székelyhidi. Note on exponential polynomials. *Pacific J. Math.*, 103(2):583–587, 1982.

L. Székelyhidi. On a theorem of Baker, Lawrence and Zorzitto. *Proc. Amer. Math. Soc.*, 84(1):95–96, 1982.

L. Székelyhidi. Remark 17. In Report of Meeting: The Twenty-second International Symposium on Functional Equations, December 16 – December 22, 1984, Oberwolfach, Germany. *Aequationes Math.*, 29(1):62–111, 1985.

L. Székelyhidi. *On the Levi-Civita functional equation*, volume 301 of *Berichte der Mathematisch-Statistischen Sektion in der Forschungsgesellschaft Joanneum [Reports of the Mathematical-Statistical Section of the Research Society Joanneum]*. Forschungszentrum Graz Mathematisch-Statistische Sektion, Graz, 1988.

L. Székelyhidi. *Convolution type functional equations on topological abelian groups*. World Scientific Publishing Co. Inc., Teaneck, NJ, 1991.

L. Székelyhidi. Ulam's problem, Hyers's solution—and to where they led. In *Functional equations and inequalities*, volume 518 of *Math. Appl.*, pages 259–285. Kluwer Acad. Publ., Dordrecht, 2000.

L. Székelyhidi. A Wiener Tauberian theorem on discrete abelian torsion groups. *Annales Acad. Paedagog. Cracoviensis*, 4:147–150, 2001.

L. Székelyhidi. On discrete spectral synthesis. In *Functional equations—results and advances*, volume 3 of *Adv. Math. (Dordr.)*, pages 263–274. Kluwer Acad. Publ., Dordrecht, 2002.

L. Székelyhidi. Functional equations on hypergroups. In *Functional equations, inequalities and applications*, pages 167–181. Kluwer Acad. Publ., Dordrecht, 2003.

L. Székelyhidi. Spectral analysis and spectral synthesis on polynomial hypergroups. *Monatsh. Math.*, 141(1):33–43, 2004.

L. Székelyhidi. *Discrete spectral synthesis and its applications*. Springer Monographs in Mathematics. Springer, Dordrecht, 2006.

L. Székelyhidi. Functional equations on Sturm-Liouville hypergroups. *Math. Pannon.*, 17(2):169–182, 2006.

L. Székelyhidi. Superstability of moment functions on hypergroups. *Nonlinear Funct. Anal. Appl.*, 11(5):815–821, 2006.

L. Székelyhidi. Spectral synthesis on multivariate polynomial hypergroups. *Monatsh. Math.*, 153(2):145–152, 2008.

R. Szwarc. A lower bound for orthogonal polynomials with an application to polynomial hypergroups. *J. Approx. Theory*, 81(1):145–150, 1995.

K. Trimèche. *Generalized wavelets and hypergroups*. Gordon and Breach Science Publishers, Amsterdam, 1997.

K. Trimèche. Wavelets on hypergroups. In *Harmonic analysis and hypergroups (Delhi, 1995)*, Trends Math., pages 183–213. Birkhäuser Boston, Boston, MA, 1998.

K. Trimèche. Convolution semigroups and Calderón's formula for compact K-variable continuous polynomial hypergroups. In *Special functions (Hong Kong, 1999)*, pages 375–393. World Sci. Publ., River Edge, NJ, 2000.

K. Trimèche. Cowling-Price and Hardy theorems on Chébli-Trimèche hypergroups. *Glob. J. Pure Appl. Math.*, 1(3):286–305, 2005.

K. Trimèche. Hypoelliptic distributions on Chébli-Trimèche hypergroups. *Glob. J. Pure Appl. Math.*, 1(3):251–271, 2005.

L. Vajday. Exponential monomials on Sturm-Liouville hypergroups. *Banach J. Math. Anal.*, 4(2):139–146, 2010.

L. Vajday. Moment property of exponential monomials on Sturm-Liouville hypergroups. *Ann. Funct. Anal.*, 1(2):57–63, 2010.

M. Vogel. Spectral synthesis on algebras of orthogonal polynomial series. *Math. Z.*, 194(1):99–116, 1987.

M. Voit. An uncertainty principle for commutative hypergroups and Gel'fand pairs. *Math. Nachr.*, 164:187–195, 1993.

M. Voit. Central limit theorems for Jacobi hypergroups. In *Applications of hypergroups and related measure algebras (Seattle, WA, 1993)*, volume 183 of *Contemp. Math.*, pages 395–411. Amer. Math. Soc., Providence, RI, 1995.

N. J. Wildberger. Hypergroups, symmetric spaces, and wrapping maps. In *Probability measures on groups and related structures, XI (Oberwolfach, 1994)*, pages 406–425. World Sci. Publ., River Edge, NJ, 1995.

N. J. Wildberger. Duality and entropy of finite commutative hypergroups and fusion rule algebras. *J. London Math. Soc. (2)*, 56(2):275–291, 1997.

N. J. Wildberger. Strong hypergroups of order three. *J. Pure Appl. Algebra*, 174(1):95–115, 2002.

H. Zeuner. The central limit theorem for Chébli-Trimèche hypergroups. *J. Theoret. Probab.*, 2(1):51–63, 1989.

H. Zeuner. Moment functions and laws of large numbers on hypergroups. *Math. Z.*, 211(3):369–407, 1992.

H. Zeuner. Invariance principles for random walks on hypergroups on $\mathbf{R}_+$ and **N**. *J. Theoret. Probab.*, 7(2):225–245, 1994.

H. Zeuner. Kolmogorov's three series theorem on one-dimensional hypergroups. In *Applications of hypergroups and related measure algebras (Seattle, WA, 1993)*, volume 183 of *Contemp. Math.*, pages 435–441. Amer. Math. Soc., Providence, RI, 1995.

H. Zeuner. Limit theorems for polynomial hypergroups in several variables. In *Probability measures on groups and related structures, XI (Oberwolfach, 1994)*, pages 426–436. World Sci. Publ., River Edge, NJ, 1995.

H. Zeuner. Lindeberg type central limit theorems on one-dimensional hypergroups. *Publ. Math. Debrecen*, 51(1-2):49–66, 1997.

H. Zeuner. A limit theorem on a family of infinite joins of hypergroups. In *Harmonic analysis and hypergroups (Delhi, 1995)*, Trends Math., pages 243–249. Birkhäuser Boston, Boston, MA, 1998.

O. Zariski and P. Samuel. *Commutative algebra. Vol. 1.* Springer-Verlag, New York, 1975. With the cooperation of I. S. Cohen, Corrected reprinting of the 1958 edition, Graduate Texts in Mathematics, No. 28.

O. Zariski and P. Samuel. *Commutative algebra. Vol. II.* Springer-Verlag, New York, 1975. Reprint of the 1960 edition, Graduate Texts in Mathematics, Vol. 29.

Index

topological vector space, 14, 96
totally bounded, 16
translate, 5
translation, ix, 5, 23
translation invariant, 96
translation operator, ix, xii, 23, 96
two-point support, 77
two-point support hypergroup, xi

unbounded, 165
uniform convergence, 22
uniform topology, 1
uniqueness, xii, 14
unit, 96
unit ball, 10
unitary equivalent, 33
unitary irreducible representation, 33

unitary representation, 33
utility theory, vii

variable, 24
variance, 27
variety, 96, 102, 109
vector space, 9

weak*-closed, 10
weak*-closure, 10
weak*-compact, 10
weak*-continuous, 10
weak*-neighbourhood, 10
weak*-topology, 10, 13, 96

Zorn's Lemma, 12